ENVIRONMENTAL IMPACTS OF RENEWABLE ENERGY

THE OECD COMPASS PROJECT

ORGANISATION FOR ECONOMIC CO-OPERATION AND DEVELOPMENT

Pursuant to article 1 of the Convention signed in Paris on 14th December, 1960, and which came into force on 30th September, 1961, the Organisation for Economic Co-operation and Development (OECD) shall promote policies designed:

- to achieve the highest sustainable economic growth and employment and a rising standard of living in Member countries, while maintaining financial stability, and thus to contribute to the development of the world economy;
- to contribute to sound economic expansion in Member as well as non-member countries in the process of economic development; and
- to contribute to the expansion of world trade on a multilateral, non-discriminatory basis in accordance with international obligations.

The original Member countries of the OECD are Austria, Belgium, Canada, Denmark, France, the Federal Republic of Germany, Greece, Iceland, Ireland, Italy, Luxembourg, the Netherlands, Norway, Portugal, Spain, Sweden, Switzerland, Turkey, the United Kingdom and the United States. The following countries acceded subsequently through accession at the dates hereafter: Japan (28th April, 1964), Finland (28th January, 1969), Australia (7th June, 1971) and New Zealand (29th May, 1973).

The Socialist Federal Republic of Yugoslavia takes part in some of the work of the OECD (agreement of 28th October, 1961).

Publié en français sous le titre:

INCIDENCES SUR L'ENVIRONNEMENT
DES ÉNERGIES RENOUVELABLES

In the context of the OECD COMPASS Project (Comparative Assessment of the Environmental Implications of Various Energy Systems) the Group on Energy and Environment of the Environment Committee has prepared this report, which attempts to assess the possible environmental impacts of renewable energy sources and systems. The International Energy Agency contributed useful comments to this work. Compared to fossil fuel and nuclear energy production and use, renewable energy sources can, in general, be considered to produce less impacts on environment and human health. It is, however, most important that their actual or potential impacts can be fully recorded and assessed so that these can be controlled and mitigated in order to ease the large development and use of these techniques.

This report was approved by the Environment Committee and derestricted for publication by the OECD Council on 18th September, 1987.

Also Available

TRANSPORT AND THE ENVIRONMENT (April 1988)
(97 88 01 1) ISBN 92-64-13045-4 130 pages £11.20 US$21.00 F95.00 DM41.00

ENERGY AND CLEAN AIR. Cost of Reducing Emissions. Summary and Analysis. Symposium Enclair' 86 (November 1987)
(97 87 07 3) ISBN 92-64-03010-7 114 pages £6.50 US$14.00 F65.00 DM28.00

OECD ENVIRONMENTAL DATA. COMPENDIUM 1987 (June 1987)
(97 87 05 3) ISBN 92-64-02960-5 366 pages £20.00 US$42.00 F200.00 DM86.00

FIGHTING NOISE. Strenghtening Noise Abatement Policies (September 1986)
(97 86 01 1) ISBN 92-64-12827-1 146 pages £7.50 US$15.00 F75.00 DM33.00

ENVIRONMENTAL EFFECTS OF AUTOMOTIVE TRANSPORT. The OECD Compass Project (October 1986)
(97 86 03 1) ISBN 92-64-12862-X 172 pages £10.00 US$20.00 F100.00 DM44.00

ENVIRONMENTAL EFFECTS OF ENERGY SYSTEMS. The OECD Compass Project (April 1983)
(97 83 03 1) ISBN 92-64-12470-5 138 pages £6.00 US$12.00 F60.00 DM27.00

IEA (International Energy Agency)

RENEWABLE SOURCES OF ENERGY. MARCH 1987 (April 1987)
(61 87 01 1) ISBN 92-64-12942- 334 pages £22.00 US$44.00 F220.00 DM98.00

ENERGY CONSERVATION IN IEA COUNTRIES (February 1987)
(61 87 01 1) ISBN 92-64-12910-3-3 260 pages £19.50 US$39.00 F195.00 DM87.00

ENERGY POLICIES AND PROGRAMMES OF IEA COUNTRIES. 1987 REVIEW
(61 88 05 1) ISBN 92-64-13110-8 494 pages £35.00 US$76.00 F300.00 DM129.00

Prices charged at the OECD Bookshop.

THE OECD CATALOGUE OF PUBLICATIONS and supplements will be sent free of charge on request addressed either to OECD Publications Service, Sales and Distribution Division, 2, rue André-Pascal, 75775 PARIS CEDEX 16, or to the OECD Distributor in your country.

TABLE OF CONTENTS

EXECUTIVE SUMMARY AND CONCLUSIONS

In general, renewable energy resources are often considered to be environmentally benign by comparison with most "conventional" energy resources like fossil fuels, which are not renewable. They are not, however, always entirely free from environmental impacts. In a sense, the impacts of the renewables may spring from the opposite qualities to those of the fossil fuels. Whereas fossil fuels are concentrated energy stores, whose rapid consumption releases into the biosphere large quantities of previously trapped elements, renewables tap more dilute energy flows which are generally of a physical rather than a chemical nature (i.e. wind, solar radiation, hot or flowing water) and therefore tend to have a physical rather than a chemical impact on the environment, e.g. noise, visual intrusion, ecosphere disruption etc. If they are to replace significant amounts of fossil fuel or nuclear capacity, renewables may require relatively larger land takes and materials inputs and their dispersed operation may sometimes pose new problems of monitoring and control.

The reader should not, indeed, be left with the impression that the balance of impacts from renewables, as described in this report, is somewhat negative. On the contrary, the important positive aspects of renewable energy use (i.e. the avoidance of environmental consequences of the production, conversion and use of an equivalent amount of energy from fossil or nuclear fuel cycles) are taken as read, and not repeated every time. Moreover, many of the impacts from renewables described here may only be encountered in rare or extreme cases, and can be the object of efficient control measures. It has also been pointed out that contrarily to fossil fuels or nuclear energy which may imply long-term and planet-wide impacts (acid precipitations, CO_2 and other greenhouse gases, radioactive-elements, etc.), renewables generally imply more localised and shorter-term effects.

Environmental impacts can take many forms. Here the possible impacts of the renewables have been explored, sometimes based necessarily on theoretical data, without attempting to rank them in order of gravity. It might be possible, in principle, to compare the noise problems of a wind energy conversion system favourably with, for example, the SO_2 output of a coal fired power plant, but in practice either may prove to be a constraint on the commercialisation of the technology. It is possible that dispersed energy installations will be better accepted by the public as, contrary to the case of more centralised systems, their energy benefits are usually reaped in the same place as their environmental penalties. In some cases, the environmental impacts, and the lifetime economic costs, will be incurred during material acquisition or manufacture of the conversion technology (e.g., silica mining for photovoltaic cells). Although this stage falls outside the traditional system boundaries, it is appropriate in the case of some of the renewables and has been analysed where relevant.

In this report, the principal technologies considered are: solar passive, solar active, solar thermal electric, solar photovoltaic, wind energy, biomass production, biomass conversion, geothermal, hydro and the various ocean energy systems. The principal environmental impacts considered are airborne, waterborne and solid wastes, water use, land and material requirements, subsidence, visual intrusion, thermal discharges, public and occupational health and safety, changes to the biosphere, and noise.

Passive Solar Architecture: These systems rely upon features of the building itself and a specially integrated architecture. Although this is generally accepted to be one of the least damaging technologies, impacts might include aesthetic considerations in building design, constraints on estate layout and design, constraints on housing densities and the removal of tree shade. Buildings might require different material inputs, including some materials that may cause problems. The possibility of lower air change rates and buildup of some air pollutants within the home has sometimes been raised. However, with new developments in air conditioning and exchange, this problem can be overcome.

Active Solar Systems (heating or cooling) are based on the use of fixed orientation solar collectors, heating water or any working fluid. They may use relatively large amounts of materials per unit of useful energy delivered, including glass, copper, aluminium, steel and insulants. Concentrating collectors delivering process heat might require large areas of land. The widespread disposal of heat transfer fluid containing biocides, rust inhibitors, dissolved metals might present problems when large numbers of systems are in use (as does the uncontrolled disposal of sump oil from cars today). Even simple panels can achieve high temperatures when drained of coolant and may cause fires, or release airborne pollutants during accidental overheating.

Thermal-Electric Solar generation plants use heliostats or parabolic concentrators to provide a receiver with fairly high temperatures. They will require larger materials input and much larger areas of land than fossil fuel plants. It has been argued that their heliostat mirrors might accidentally cause temporary or permanent blindness.

Photovoltaics transform directly the photons from incident sunlight into electrical current through the use of semi-conductors. Although benign in use, they require quite exotic inputs, some of which are explosive (silane), toxic (cadmium sulphide), or produce toxic substances during refining (mercuric effluents from gallium extraction). Photovoltaic cells may cause fires and may release toxic substances during combustion. Their environmentally sound disposal at the end of their useful life is also a factor.

Wind Energy conversion systems require windy, exposed sites and may therefore present a significant visual intrusion, especially when constructed in large clusters. They can cause local disruption to electromagnetic communications, including TV reception and radio transmission. Aerodynamic noise from the rotors has been experienced in a number of cases and may prove to be a constraint on future developments of large-scale units. Rotors can disintegrate, throwing blades over a short distance, or can shed ice during normal operation.

Using Biomass from a large number of sources can have complex environmental consequences. The use of those sources involving urban or industrial wastes, sewage sludge and animal wastes may offer some environmental benefit. However, the removal of crop or tree residues for energy production could gradually cause soil degradation (lowering of soil organic content) and encourage soil erosion or leaching of synthetic fertilizers. The use of land specifically for energy crop production might require large inputs of water, synthetic fertilizers, pesticides, etc. (with the usual consequences), and would remove land from normal agricultural uses and food production. The wide range of environmental problems associated with monocrop agriculture may be encountered on a large scale, in energy plantations.

The conversion and end-use of biomass can release a wide range of airborne, liquid and solid wastes. Direct combustion in home units is rather polluting, releasing quantities of particulates, CO, toxic organic substances, and NO_x. Burning urban or industrial waste may be environmentally hazardous, as the waste stream may give rise to toxic substances such as chlorinated compounds (dioxine for example). Other conversion processes may release H_2S, ethanol, and a range of complex organic substances. Many organic waste products can also be used as fertilizers, chemical feedstocks, etc. If dumped, however, they could constitute a major water pollutant, due to their biochemical oxygen demand, nitrate and phosphate content, etc.

Geothermal Heat is effectively "mined", and produces a wide range of airborne, and liquid wastes. These vary considerably from site to site but include: dissolved salts, dissolved metals, gases including H_2S, CO_2, NH_3, mercury and arsenic which can be released either in liquid wastes or in the stream. Steam systems produce noise at high volumes, especially during development and early production. Water dominated systems can cause subsidence.

Hydroelectric generation, which is generally based on the construction of a dam and reservoir, can have both positive and negative specific effects on the environment. The positive effects in particular are: flood control, flow regulation (increasing low flow and water supplies during dry seasons), and the recreational use of the lake. Negative aspects include: the possible loss of large areas of farmland and human settlements; disruption of fish cycles (migration and reproduction); disruption in sediment and silt transportation by the river, with increased erosion downstream and other impacts; alteration of water quality (due to eutrophication); the possibility of dam failures. These positive and negative aspects are generally mixed and their balance will vary for each individual case. The new projects carried out within a multipurpose prospect have generally a better environmental and socio-economic balance than those created purely for energy production.

Ocean Energy harnessing has given rise to a relatively large number of technologies using different approaches (tidal energy, wave power, ocean thermal energy conversion). The environmental impacts are mainly on fish and aquatic life, navigation and recreation. Tidal energy schemes, when situated on estuaries, have probably the most noticeable effects in this respect, which are comparable to those of classical hydropower schemes, although more complex, as they may cumulate impacts on both fresh and sea water (particularly fish reproduction and migration).

For the moment, large-scale exploitation of renewable energy sources has not started (except for hydropower and fire wood). The present analysis of possible effects is therefore rather "prospective" as real experience is still often insufficient. It is thus essential that identification and anticipation of potential impacts, prior to any large-scale developments and implementation, can be carefully taken into account so as to enable preventive measures to be better integrated into systems which are not yet in substantial commercial use (thereby to avoid much of the damage which was sustained by human health and the environment due to the exploitation in times past of conventional energy forms, i.e. coal).

Large-scale development of renewable energy sources, especially medium- and long-term, is necessary and environmentally desirable. Serious environmental issues, such as atmospheric CO_2 build-up and climate change, may require a substantial reduction in the use of fossil fuels in the future (especially coal, which produces more CO_2 per energy unit) and their replacement by the use of appropriate renewable energies.

Chapter 1

RENEWABLE SOURCES WITHIN A GLOBAL ENERGY PROSPECT

INTRODUCTION

Faced with the inevitability of an accelerating decline in reserves of low-cost fossil fuels, energy planners in Member countries have been giving increased attention to the prospect of developing alternative supplies based on natural energy flows. The technologies associated with this approach have been given a variety of names over the past few years -- "alternative", "soft", "ambient", or simply "new". The most widely adopted generic title is now "renewable" and this term will be used throughout the report. The term renewable reflects the idea of exploiting energy flows that are continuous, in contrast to most existing supply technologies which exploit energy reserves built up over long geological periods (such as fossil fuels). Almost all the renewables involve direct or indirect harnessing of solar energy; however, tidal power exploits the rotational energy of the earth and moon. At the margin the definition is extended to embrace sources that exploit short-term energy stores (e.g. standing biomass) or very large natural heat stores (geothermal). Although the latter may not be, strictly speaking, a renewable source, it will be considered here as it is often treated as being part of the renewable group.

Most of the renewable energy sources currently being developed have been tapped before -- sometimes over long periods, though past applications were often rather rudimentary and on a small scale. They were used to deliver mechanical power or heat. During the second and third quarters of the twentieth century these technologies were largely abandoned in favour of energy systems based on cheap fossil fuels. Those now under development are in some cases large-scale and most of them deliver final energy in the form of electricity or heat (Table 1.1). Operating experience with the majority of these technologies is still sometimes limited but progresses rapidly..

The period of fossil fuel domination has also seen a considerable increase in energy consumption and a substantial expansion in the ownership of fuel-using appliances. It was during this period that Member countries created, or greatly augmented, their present energy infrastructures. Energy supply has become almost exclusively associated with large, centralised production facilities and extensive integrated transmission grids. In the period following the Second World War, coal and coal-derived solid fuels have tended to be displaced from the rapidly-growing domestic fuel market. In part

Table 1.1

RENEWABLE ENERGY TECHNOLOGY DEVELOPMENT STATUS IN APPLICATION SECTORS(1)
(Worldwide Overview)

Conversion technology	Hot water	Space heating	Cooling	Process heat	Electricity	Mechanical uses	Liquid fuels	Gaseous fuels	Solid fuels
	Development status Application/end use								
	Direct Utilisation of Solar Radiation								
Active solar technology									
Flat plate collector (fluid heat transfer)	xxxx	xxx	xxx						
Air collector systems		xxxx	xx	xx					
Evacuated tube collector	xxxx	xxx	xx	xx		xx			
Passive solar technology									
Solar thermal technology									
Concentrating collectors			xx	xxx	xx				
Heliostat systems				xx	xx				
Photovoltaic									
Small systems (kW-range)					xxxx	xxx			
Large systems (MW-range)					xxx				
Space station					xxx				
	Indirect Utilisation of Solar Radiation								
Hydropower utilisation									
Large systems					xxxx				
Small systems					xxxx	xxxx			
Wind energy utilisation									
Small systems (50 kW)		xx			xxxx	xxxx			
Medium-size systems (500 kW)					xxx				
Large systems (1 MW)					xxx				
Bioenergy utilisation									
Thermochemical conversion									
Combustion-small (kW-range)		xxxx							
-large (MW-range)		xxxx		xxxx	xxxx				
Gasification					xxx			xxx	
Pyrolysis/synthetic gas							xx	xx	xxxx
Liquefaction							xx	xx	xxx
Biological Conversion									
Fermentation							xxxx		
Anaerobic digestion-small		xx		xx	xx			xxxx	
-large		xx		xxx	xxx			xxx	
Ocean energy									
Tidal					xxx				
Wave					xx				
Thermal gradient					xx				
Geothermal utilisation									
Geothermal anomaly		xxxx		xx	xxxx				
Hot dry rock (HDR)		xx		xx	x				

x = Basic R&D
xx = Experimental/pilot system
xxx = Full-scale demonstration
xxxx = Commercial availability

this has reflected non-price factors -- notably cleanliness, ease of use and the difficulties of solid-fuel storage -- and in some urban areas at least the establishment of strict air- pollution controls, has helped to diminish the attractiveness of solid fuels. One result of this process has been the establishment of a widely-dispersed electricity-specific demand throughout all market sectors.

ENVIRONMENTAL IMPACTS

The total environmental impact of the renewables will depend on which particular technologies are favoured, how large a part of overall energy supply they are expected to meet and on the way in which they are deployed. Proponents of renewable technologies argue that their benefits flow in part from their ability to deliver appropriate energy (e.g. low-temperature heat for domestic space-heating) in a dispersed system, independent of large utilities. However, the existence of well-developed large-scale electricity grids and the existing institutional framework of energy supply may tend to favour renewable technologies that share the characteristics of thermal power stations.

In most cases, the environmental impact of these technologies will depend on the scale and the geographical distribution as well as inherent properties of the energy source. This analysis has therefore, where relevant, taken account of different options within a single energy source.

THE ENERGY CYCLE

It is impossible to set a definitive boundary on an energy system. In considering the plausible range of environmental impacts arising from the adoption of a particular technology, a truly rigorous analysis would need to consider all those economic and social activities initiated by a particular technological choice. At present the methodological tools to enable this to be done properly are not available (2). Even if large-scale input/output models of entire economies could be constructed, they would not allow the researcher or policymaker to avoid a number of fairly arbitrary choices about system boundaries, particularly where resources are only partly used for energy purposes (e.g., dams with electricity generation, railroads for bulk coal transport).

This poses particular problems in the case of those renewable technologies that are widely dispersed. In many respects, the systems needed to provide useful energy from fossil and nuclear fuels are, though different in detail, broadly comparable. Their environmental impacts largely flow from the same fundamental properties. Fossil fuels represent very compact energy stores. Their use therefore involves impacts arising from their geographical and geological remoteness (mining, exploration, transport) and the release into the biosphere of large quantities of compounds (e.g. CO_2, SO_2 and NO_x) in a short space of time and at concentrated locations. Their energy intensiveness makes them liable to catastrophic releases: gas explosions, nuclear reactor incidents or oil field fires. At the same time, it should be recognised that centralised energy systems offer certain advantages. In some cases, remedial technologies are available, but at a cost (e.g. flue gas desulphurisation). In other cases more environmentally favourable technologies are themselves more likely to be adopted (e.g. fluidised bed combustion). Institutional control of environmental damages, including better management techniques and the establishment of external monitoring, may also be easier to introduce in centralised systems.

As we know, renewable sources generally convert relatively dilute energy flows. Therefore, comparatively large amounts of conversion hardware are needed to produce useful energy. Furthermore, their application is likely to require large resource inputs in terms of both land and materials, and the control of their performance over time will need new maintenance and surveyance capabilities. In some cases, the materials involved are relatively scarce (silver) or have potentially serious environmental consequences (fluorocarbons, gallium arsenide). In this sense, the potential danger is precisely the opposite of that with fossils -- the substances, rather than being consumed rapidly, will be stored for quite long periods in a large number of locations and, possibly, disposed of at the end of their useful life in a manner that is hard to control. For example, active solar panels may use heat-transfer fluids that require flushing and replacing every two or three years. The fluids, even if water-based, may contain anti-freeze, rust inhibitors, and biocides which can be hazardous to the environment. The problems associated with their uncontrolled release into the biosphere are foreshadowed by the disposal of sump oil and cooling water from motor vehicles.

Most environmental impact appraisals of conventional energy supply systems have taken the energy cycle from the extraction of the primary fuel to the production of useful energy -- in the case of coal, for example, this runs from mining to benefication, to conversion, to transmission, to consumption, to production of useful energy. It is not usual to consider, for example, the environmental impacts of manufacturing mining equipment or nuclear installations (though on-site construction disruptions may well be considered). However, in the case of many of the renewable technologies a significant part of the overall environmental risk -- and indeed of the lifetime economic cost -- may be incurred at the manufacturing stage. This can produce distortions in making comparisons with other conventional energy systems and may give the impression that renewable sources have more environmental impacts than they have in reality.

Equally, most analyses of non-renewable energy sources fail to consider the consequences of fuel-use on the internal environment in buildings where the fuel is finally used, though a limited literature now exists on indoor pollutants. Human intoxication due to gas leaks, or to CO releases in the case of coal burning stoves, are well known. Decentralised renewable energy technologies tend to be closely linked to substantial demand-reduction through increased insulation, either as an explicit policy goal or because the technology delivers limited and/or fluctuating supplies of heat. It has been argued that this might lead to increased condensation or concentration of indoor pollutants. These problems are considered in the section on solar architecture, though they may well apply to a wide range of low-energy buildings. On the other hand, some of the renewables produce conventional end-use energy and fuels (electricity or methane). The environmental implications of these fuels has been explored elsewhere in the COMPASS programme and will not be considered in depth here (65).

The relationship to other energy production, conversion and delivery systems is clearly an important part of any analysis of the impacts of renewable technologies. In net system terms, the overall impact needs to take account of displacement, as it can be roughly assumed that the addition of a megawatt of, for example, hydro-electric capacity could remove the need for a megawatt of thermal capacity. In practice, the balance may be extremely complicated, especially if a dispersed system delivering low-temperature heat is developed within an existing centralised network delivering electricity. Nevertheless, it is important to consider the impacts across the whole cycle -- including the effect on load factors, competition for sites, additions to or subtractions from the transmission and distribution network and the technology associated with energy storage and transportation. Displacement effects might go beyond the energy economy. For example, the conversion of urban refuse or agricultural wastes into fuels might avoid the need for traditional disposal methods that may be themselves environmentally damaging.

ENVIRONMENTAL ASSESSMENT

Environmental impacts from energy use range from the purely local (noise from a geothermal well) to the global (CO_2 release from fossil-fuel burning). On the whole, the impacts of the renewable sources are localised. Most notably, the renewables tend not to produce large quantities of airborne sulphur oxides (linked with widespread acid deposition) or of carbon dioxide. Indeed, operating solar, wind and hydro technologies produce virtually no airborne pollutants at all. Renewables are widely perceived, therefore, as being "clean" technologies. Nevertheless, even the cleanest of the renewables may have some impact on the environment and these may, unless anticipated, constrain the penetration of these energy services. The willingness of populations to tolerate degradation of the environment varies from place to place and from time to time. Unforeseen environmental penalties in new technologies, however insignificant they may seem to the designers, can be a significant barrier to their commercialisation. The variation in local resistance to new energy developments in general is essentially a sociological question, though it is arguable that it arises, at least in part, from the concentration of environmental sacrifices in particular localities while the benefits are enjoyed elsewhere. To a large extent, the renewable technologies can avoid this problem (though it does not follow that they will) as most are capable of being developed in a dispersed fashion, without incurring significant diseconomies.

The environmental impacts considered here include: water use; subsidence; material use; heat pollution; visual intrusion; population shifts; land-use conflicts; solid waste disposal; air and water emissions; public and occupational health and safety; changes to natural habitats; noise.

THE POTENTIAL CONTRIBUTION OF RENEWABLES IN ENERGY SUPPLY

The scale of the likely environmental impact of renewables will depend on the contribution that the technologies are expected to make, and on whether dispersed or centralised systems are adopted.

Hydroelectric Plants are compatible with centralised electricity networks and are competitive with thermal generating capacity. They have, therefore, established themselves as a commercial technology in all OECD countries (with the sole exception of the Netherlands) and account for a very large proportion of electricity supply in some. Between 1960 and 1982 hydroelectricity production in OECD more than doubled. The rate of increase failed to keep pace with the overall increase in electricity generation, however, so the proportion of electricity produced by hydro fell from 32 per cent to 21 per cent. Hydro retained its share in some countries, for example, in Norway over 99 per cent of electricity was generated by hydro throughout this period, although Norway's overall electricity production rose almost as quickly as that in OECD as a whole. (3)

Hydro is the only renewable technology that is already established as a major contributor to OECD energy production. Total installed capacity has been estimated at 285 GWe (4); an additional 60 GWe is under construction. This represents a significant proportion of the available large-scale hydro resources in Member countries -- about 200 GWe of potential is thought to remain. The development of hydro to date has largely consisted of exploiting the larger sites. Now that many of these have been developed, interest is turning to the potential for small hydro power (SHP). SHP is not precisely defined, but following the IEA (4), is taken to mean systems of up to 10 MWe in size. The IEA has suggested a definition in which stations of under 100 kw capacity are designated as "micro", those between 100 and 1 000 kw as "mini" and all stations up to 10 MW as "small" (4): other analyses use other limits. The IEA estimates that the inclusion of SHP sites adds a further 15 or 20 per cent to the remaining potential capacity. A report prepared for the U.K. Department of Energy (5) examining the potential for SHP development (defined as sites offering power output greater than 25 kw) potential, notes that "...as the capacity limit is lowered, the number of sites increases exponentially". It seems likely that SHP will make a significant contribution to the expected increase in installed hydroelectric plant (HEP) capacity, used either to feed electricity into national grids, or to meet local needs in areas remote from transmission networks.

Biomass -- in particular firewood -- already plays a significant part in overall energy supply. It is estimated that in 1978 biomass accounted for 5.4 per cent of world commercial energy, though figures of this sort need to be treated with caution as only a small proportion of the biomass used will pass through the commercial energy system. Its role is most important in developing countries, where it accounts for at least 20 per cent of energy use. Within the developed world it probably accounts for less than 1 per cent, though there is considerable variation between OECD countries as it still supplies over 10 per cent of total primary energy in Turkey and Portugal and 7-8 per cent in Sweden (6).

Geothermal Heat is estimated to supply about 100 000 homes and commercial buildings. Twelve countries use direct geothermal heat on a significant scale (including seven OECD members). The total heat used in this way has been estimated at about 2700 MWth (6, 7). It is also possible to generate electricity from geothermal heat. Twelve countries (including seven OECD) use geothermal generating sets, producing a total of about 2500 MWe of capacity. It has been calculated that this figure could rise more than sevenfold by the end of the century (7), though even then it will only supply a small part (not more than 1 per cent) of overall energy demand. Recent IEA estimates (for OECD countries) put the possible contribution at 15 GWe in 2000 (8). (This figure does not include Iceland and France.)

Direct Solar is already in widespread production (Table 1.2). It is believed that there are around 600 individual manufacturers operating within OECD, and it is estimated that about 10^6 m^2 of panels have been installed in OECD countries. Much of this is dedicated to the collection of low grade heat, notably for use in swimming pools or for domestic hot water. The contribution of direct solar through the rest of the century is likely to come largely from dispersed uses. These are particularly difficult to quantify, as they depend on the decisions of millions of individual consumers. Recent forecasts reflect this uncertainty.

Table 1.2

PROJECTION OF SOLAR CONTRIBUTION TO
U.S. PRIMARY ENERGY IN THE YEAR 2000 (MTOE)(8)

Year projection made	Source of projection	Passive	Active
1979	Domestic Policy Review: Maximum Incentive Option	25.2	50.4
1981	NEPP	11.3	6.3
1981	EIA Annual Report to Congress mid case	4	15.1
1982	Journal of Energy	25.2-50.4	50.4
1983	NEPP Scenario B	10.1	5.0

Photovoltaic worldwide production capacity in 1983 has been calculated at 22 MWp. At present, the principal commercial uses are in very small-scale consumer applications (watches and calculators) where price is not a critical determinant, or in technical non-grid connected applications (communications, telemetry, etc.) where the price is already competitive. It is anticipated that more mainstream applications will start to become commercial by the beginning of the 1990s, particularly stand-alone uses, including village power systems. This will account for the majority of the growth during the latter part of the century, though grid-linked systems are expected to start becoming important in the early 21st century. Projections of total world markets in 1990 give a range of between 100 and 500 MWp installed, with the best estimate between 350 and 400 MWp. Projections for installed capacity in 2000 range from 15 000 to 40 000 MWp. Much of this growth is expected to occur in the less developed countries and OECD uptake might be slower than these figures suggest, for example, photovoltaic does not start to make a significant contribution in the U.S. National Energy Policy Plan projections until 2005 (9).

Wind Power is being developed differently in different countries, some opting for the development of many small systems, some for a restricted number of large devices. The Danish projection of the contribution made by wind to its primary energy needs in the 1990s anticipates that 1 400 GWh of electricity will require 37 000 individual machines, whereas Norway foresees the generation of 1 000 GWh in the year 2000 from only 35 units. The IEA estimate that the total combination of wind power will rise from 1 526 GWh in 1985 to 194 680 GWh in 2000, though this figure is dominated by the anticipated US contribution of 170 000 GWh, involving 60 000 WECS. The total contribution in all nine countries is forecast to involve the use of over 130 000 machines.

Renewable energy sources are going to become increasingly important in the years and decades to come. Apart from hydropower and some biomass technologies which are quite classical, the other renewable energy technologies are a large and complex family in constant evolution and development, and break-throughs can, within a relatively short time, upgrade the energy value and economic competitivity of specific techniques. For the moment, a certain ranking may be given (8) to these different techniques. In the category A/B are those technologies which are well developed, commercially available and economically viable, although some of them may need favourable conditions and preferential treatment (i.e. tax incentives) to be competitive. They include:

-- Solar water and space heating in general;

-- Small photovoltaic systems in remote locations;

-- Small wind systems;

-- Conventional geothermal technologies;

-- Tidal systems;

-- Biomass combustion (power and heat) and biomass liquid and gaseous fuels.

In category C/D we find those technologies which still need more development to improve their efficiency, reliability and cost, so that they can become really commercial and competitive within a broader range of conditions. Governments should support efforts in this direction, but some of these technologies are already commercially available and could be usefully put into operation, advantage being taken of favourable geographical and economic conditions:

- -- Solar thermal power systems and photovoltaic power systems;
- -- Ground space cooling;
- -- Biomass energy crops;
- -- Medium/large wind power systems;
- -- Wave energy systems;
- -- Ocean thermal energy conversion systems; and
- -- Various unconventional goethermal systems (binary, magma, goepressured, hot dry rock).

The majority of renewable sources and technologies transform the initial energy into electrical power. Today, in many OECD countries, electrical utilities often have a status of monopoly, which makes it somewhat dissuasive for individuals, municipalities or enterprises to consider the adoption of renewable techniques for their supply. However, from the viewpoint of utilities, it is more economic and makes more sense in general to concentrate investments on large (fossil, nuclear, hydro) power plants than on dispersed renewable sources of a small/medium dimension. The renewable energy sources at the present time are, indeed, more promising for decentralised use like solar heating, single house photovoltaic and small wind systems, than for large centralised units. This is the role of governments to assess, orient and promote the different renewable technologies in the specific fields where they can be competitive, and to provide them with the appropriate support, if they wish this type of energy to make a real start.

Chapter 2

ENVIRONMENTAL IMPACTS OF RENEWABLE TECHNOLOGIES

SOLAR ENERGY

I. SOLAR ARCHITECTURE

Solar architecture is a passive solar technique in current production (Table 2.1).

1. Orientation and Appearance

Passive built solar buildings take a variety of forms but certain characteristics are common to all. Passive solar buildings require careful site orientation and fenestration. Up to 90 per cent of the total window area is placed on a single wall of the building (the Southern wall in Northern-hemisphere applications) Ideally, all buildings should face south, though the precise orientation can vary by up to 30 per cent without significant loss of efficiency. The performance of the building can be influenced by careful control of the micro-climate, and this could lead to a range of design measures aimed at controlling surface areas (by building in terraces) and wind velocities (by building walls, tree-planting etc). An unavoidable requirement is that the buildings should not be shaded: some sites may be totally unsuitable due to the presence of tall buildings; others may require the removal of existing trees.

The degree to which the visual appearance of the houses might need to diverge from local traditions will be in part a product of the level of energy self-sufficiency. However, present knowledge and experience suggests that designs that are intended to achieve significant (50 per cent or more) reductions in conventional energy requirements are likely to be dictated largely by energy considerations. Architectural aesthetics are inherently subjective, but it is possible that the resulting designs might be considered too "radical" if sited among existing buildings of more traditional design, or too repetitive if proposed in large numbers on a greenfield site. The reaction of local planning authorities and existing residents might, in some cases, prove to be a constraint on a large-scale adoption of passive solar designs, (i.e., in cities and regions of historic and traditional habitat).

Table 2.1

SOLAR TECHNOLOGIES: TECHNOLOGY REVIEW

Technology	Delivered Energy	Scale	Examples
Passive Architecture	Warm air Heat 30°C	Building or estate	Milton Keynes Pennylands
Solar Pond (Planned)	District heat 90°C Electric 150Kw Electric 5Mw	Small to 7 000m^2	Israel Israel Salton Sea (Calif.)
	ACTIVE SYSTEMS		
Black panel Vacuum tube Line focus Parabolic Segmented Fresnel	Heat 90°C Heat 90-300°C Electric	Individual building or estate	Around 600 manufacturers in OECD About 10^{6}m^2 installed in OECD
Solar farms	Electric 1Mw		Nio (Japan)
	Electric 500Kw	100 acres	Almeria (Spain)
Central receiver	Electric 10 Mw	9 000m^2 heliostats	Barstow (Calif.)

2. Materials

Solar buildings need to be designed to much higher heat-loss standards than are presently common. The reduction of fabric heat-losses is achieved by increasing thermal insulation levels. This requires a greater use of insulating materials than in conventional housing, imposing some penalties both at the construction and the demolition stage. Although most of the materials will be perfectly commonplace, it is not unknown for materials connected with insulation or heating systems to prove hazardous to human health (e.g. asbestos, urea formaldehyde (UF) foam).

Materials used include:

Cavity wall fill: urea formaldehyde foam; expanded polystyrene blocks or beads plus bonding agent; expanded mineral wool, etc.;

Roof space: glass fibre; cellulose wool; polystyrene beads; metal foil etc.;

Window area: multiple glazing; shutters or insulating curtains;

Floors: layer of plaster, expanded polystyrene etc.

3. Air Change Rates

Heat exchanges with the surrounding atmosphere also need to be controlled. This does not neccessarily imply reduced air change rates. Heat-exchange systems can decouple heat-loss and air-change rates by recovering useful heat from the evacuated air. However, these add to the cost of building and it is to be expected that air change rates in energy-conscious building designs may be somewhat lower than in conventional buildings.

There also needs to be some method of minimising diurnal temperature variations, either by providing a heat store and/or by designing a slow thermal response into the building. As communal heat-stores require large areas of land (unless built underground), their use may be unlikely in residential or commercial areas where land is at a premium. Internal storage in some present designs involves the retention of bodies of warm air in roof spaces or dry rock stores. Air-circulation patterns are different from those in conventional buildings.

Low air-change rates and unusual patterns of air circulation might produce a range of environmental problems within the building. Air changes carry evaporated water out of the building and the retention of saturated air can lead to humidity/condensation problems. Some existing pollutants may also concentrate within the internal house environment. Pollutants can for example be produced from a number of sources:

-- Carbon monoxide is a toxic gas released by the combustion of most fuels in conditions of oxygen shortage (i.e. faulty heating appliances). Reduction of ventilation in rooms where fuel burning appliances are used, can cause the build-up of CO. It is difficult to detect, being odourless and colourless and the symptoms of CO poisoning are vague, often resembling a cold or flu.

-- In general oxides of nitrogen can be produced by heating or cooking appliances in homes that use gas (10) and an unflued (or inadequately flued) fossil fuel appliance. NO_x has been associated with respiratory illnesses, particularly in children, though there is not at present sufficient evidence to conclude that internal dwelling exposures can cause respiratory or other problems.

-- Radon is a radioactive gas released by building materials (i.e. granite) and by subjacent ground. The geology underlying a building is the dominant factor. Health effects arise from the inhalation of radon and its decay products. Various attempts have been made to estimate the overall health effects, though these remain controversial.

-- Formaldehyde can be produced by Urea Formaldehyde foam, sometimes used as an insulator (has now been banned in a number of countries). It is also given off by a range of products used within the home including some cleaning materials. Formaldehyde causes skin and mucous membrane irritation in humans. Longer exposures can cause vomiting, diarrhea, coughing, etc. Formaldehyde has induced squamous cell cancer in rats, though its carcinogenic effect on humans is still in doubt.

-- Asbestos may be present in the house in old building materials or heating appliances. If abraided it can produce airborne fibres that can lodge in the lung walls causing long-term damage (cancers for example).

It is important to note that such impacts are, of course, not linked to the use of renewable energy sources, but only to building devices where aeration is badly designed.

4. System Effects

Passive solar architecture may require no additional energy conversion or transmission facilities, though the majority of solar homes will continue to be connected to main fuel grids in the same way as normal buildings, certainly for electricity-specific demand and probably for backup space- and water-heating. The percentage of the building's total energy requirements supplied by solar gain will depend on the local climate, the design and siting of the building and the behaviour of the building users. In a number of existing examples (St. George's School, Merseyside, U.K.) the energy efficiency of the building is great enough to obviate the need for auxiliary heating fuels. The U.S. Department of Energy reference design (11) shows that as much as three quarters of the building's energy needs can be satisfied by passive solar heat. Where additional fuel is needed, it is likely to be supplied by conventional sources.

Two possible effects of the backup systems can be predicted. On the energy supply side, the reduction in demand for fossil fuels should be environmentally beneficial. However, the precise effect on the central electricity network of a large takeup of passive solar architecture is likely to prove more complex than this. Although average demand over the year will be reduced, the structure of the diurnal and yearly demand curves will also be distorted, producing changess in the use of generating plant and affecting the economics of electricity production. Major utilities within Member countries should be able to elucidate the system effects of increasing the availability of solar (and indeed of other non-firm renewable energy sources) on the use of conventional generation capacity.

At the consumer end, the reduction in heating loads might produce a change in fuel preference. The cost to the conusmer of piped or wired fuels normally reflects three costs: fuel and capital costs, peak demand costs and the fixed costs of providing a supply. This last includes the cost of local distribution networks, emergency backup services, the supply of an individual meter, meter reading and billing costs etc. Normally, these costs are recovered through standing charges in a two-part tariff. When the fuel use of individual households is large, this element represents a fairly small proportion of each bill. However, the costs to the utility remain the same however much or little fuel is supplied. If households (or commercial premises) succeed in drastically reducing their demand for piped fuel these costs will start to dominate bills. There will therefore be an incentive for low-consumption customers to switch away from piped or wired supplies altogether. In some (particularly rural) areas, for example, there may be an increased use of firewood to meet auxiliary fuel needs (see Biomass Section).

5. Land Use

Passive solar architecture should impose no additional land use requirements. Net land savings should therefore be possible when the reduced requirements for energy generation and transmission are taken into account. It is also worth nothing that most of the environmental impacts are experienced at the same location as the benefits of the energy use. In principle this should mean that some of the environmental costs will be "internalised".

6. Conclusions

The large-scale introduction (or the retrofitting) of solar architecture in areas where regional styles and traditional architecture are an important element for cities and landscapes may pose obvious problems unless special care is taken to adapt technologies and materials to existing styles. The possibility of reduced air change rates and accumulation of some indoor pollutants has also been mentioned, although these are not specific to solar architecture, and normally should not be a problem.

II. SOLAR PONDS

Solar ponds are among the cheapest available solar technologies, and require almost no materials input, apart from the pond linings and the salts themselves. Their use of land, on the other hand, may be extensive, especially if the intention is to use the heat generated in a thermal electricity system. The ponds are unlikely to have any secondary benefits, compared, for example, to the irrigation, fishing and leisure uses to which large hydro lakes can be put. The water in the ponds is heavily saline and will normally be treated with biocides to prevent algae or weed growth near the surface. Large-scale leaks could lead to the contamination of nearby land and aquifers. As the ponds produce quite low-temperature heat, their use in electricity generation will require, for example, organic rankine cycle engines that might in turn involve light environmental risks.

III. DISPERSED ACTIVE SYSTEMS

Although a range of different systems exist, their environmental impacts differ mainly in degree, and can be treated together here.

1. Materials

Flat Plate Collectors consist of: absorber plate -- usually copper, steel or aluminium; selective surfaces; heat transfer fluid (indirect systems) -- usually water and range of additives; heat transfer pipes -- usually copper; insulation; glass or clear plastic cover; heat sensors; pumps (active systems).

Total input of materials for two reference active systems, one residential and one commercial, based in Denver, Col., were calculated by the US DOE (11). Table 2.2 provides an estimation of principal materials required to produce flat plate collectors with a total energy output of 10^{15} J per year.

Vacuum-tube collectors use similar materials to flat plate collectors though their greater efficiency enables them to achieve higher temperatures (depending on the heat transfer fluid used) and to use less materials.

Table 2.2

ACTIVE SOLAR MATERIAL USE PER 10^{15} J PRODUCED (MT)

	Residential	Commercial
Fibreglass	3 500	29 000
Steel	9 300	49 000
Glass	4 900	49 000
Copper	1 800	10 000
Aluminium	6 800	68 000

Although many of the solar collectors currently use quite large quantities of materials, it must be remembered that this is a once-for-all economic or environmental cost repeated only at the end of the life of a unit (currently estimated at 20 years). It should also be borne in mind that the material input should properly be compared to the construction inputs of the conversion and transport technologies associated with conventional fuels -- mining machinery, drilling rigs, coal trains, gas compressor stations, pipelines and grids. This sort of analysis is inevitably more complex but should be done if systems boundaries are to be kept consistent.

2. Land Use

Low/Medium Heat: Land use depends on the system chosen. In the case of single-dwelling hot water or space heating/cooling, the system will usually be added to the roof of the existing building, and no land will be required. Communal low-temperature systems might use some land, though again the collection surfaces might well be added to already existing buildings. The principal additional use of land might be for heat storage.

High Temperature: The land-use requirements of concentrating collectors providing process heat are more problematic. A reference system used by US DOE employed 47 m^2 of collectors to provide 65 x 10^9 Joules. The actual land-use/collector-surface ratio of 1.3 used by DOE indicates that 60 m^2 of land were used to provide this amount of energy.

3. Routine Discharges

Once operational, solar collectors require very little material or water throughput. Coolant liquids may need changing every three years or so. In the US DOE reference models, the total throughput of coolant per 10^{15} Joules of energy produced was between 750 and 2 700 thousand liters. Discharges like these may require careful disposal. In some cases, the coolant will be water based, but all indirect systems are likely to contain anti-freeze, rust inhibitors or biological growth inhibitors, as well as substances leached from the system during use. Heat transfer fluids might therefore contain glycol, nitrates, nitrites, chromates, sulphites, sulphates. Higher temperature applications would use more exotic substances -- aromatic alcohols, oils, fluorocarbons or liquid sodium. The large-scale adoption of solar technologies might well require some controls to be placed on the way that these are dumped.

4. Accidental Discharges

Apart from normal use, there may be the danger (especially with domestic hot water systems) of accidental pollution of potable water supplies through leaks of heat transfer fluid.

Solar converters can achieve relatively high temperatures if their coolant is lost. Conventional flat panels with selective surfaces can reach 200°C when dry. At this temperature there is a risk of fire, with the additional problem of outgassing from panel components (insulant, plastic components, epoxys) and the release of heat transfer fluids in gaseous state or following combustion (e.g. from burnt freon).

5. Impact on Building Stock

The aesthetic impact of solar panels is evidently a matter of taste, though it can be argued that their effect is no more distasteful than television aerials, which are widely accepted. Flat panels can be designed in a way that fit closely with the existing roof line and produce little glare. The addition of such equipment to the building fabric may increase the risk of fire or water intrusion into the roof space.

6. Conclusions

Although the manufacture of active solar systems requires relatively large quantity of materials on a per unit of energy output basis, very little material is consumed once the systems are operational. The only potential source of environmental pollutant during the systems operation arise from the coolant change, which can be easily controlled by secured disposal. An accidental leakage of coolant in the systems, however, can cause fire and gas releases from vapourised coolant, adversely affecting public health and safety. On the other hand, the large scale adoption of active solar technologies will significantly reduce the burning of conventional fuels and consequently, reduce the environmental impact associated with these fuels. This favourable contribution to the environment by active solar systems is implicit and has not been assessed in detail in this study.

IV. SOLAR THERMAL CENTRAL-STATION ELECTRICITY PLANTS

1. Introduction

In a solar thermal central-station electricity plant, heat energy from the sun is converted into mechanical energy in a turbine and finally into electrical energy by means of a conventional generator coupled to a turbine. The efficient conversion of heat energy into mechanical energy and hence electricity, requires the working fluids supplied to the turbine to be at a high temperature (i.e., above 175°C). To obtain such temperatures from a solar source requires the use of focusing concentrating collectors. Two basic arrangements have been proposed for converting solar radiation into electrical energy: i) the central receiver system; and ii) the distributed collector system.

In the central receiver system or the "power tower" design, an array of sun-tracking mirrors (heliostats) reflect solar radiation onto a receiver mounted on top of a central tower. Solar energy absorbed in the central receiver is removed as heat by means of a heat-transport fluid and converted into electrical energy in a turbine-generator.

The distributed collector system may consist of a number of parabolic trough-type (line-focusing) collectors or of paraboloid dish-type (point-focusing) collectors. The absorber pipers (or receivers) of the individual collectors are connected so that all the heated fluid is carried to a single location where the electricity is generated.

The basic difference between the central receiver and distributed collector systems is that in the former the solar energy falling on a large area is transmitted to a central point as radiation, but in the latter, the energy is carried as heat in a fluid. Analysis of these two systems indicates that they may have different preferred applications. The central receiver system appears to be more suitable for the large-scale power generation, while the distributed collector design appears to be more suitable for small power facilities less than 2 MW in size. Of these two alternatives, the central-station design has received most attention in the U.S. (12).

2. Technology Review (12)

In the USA, the first central receiver system built for conversion of solar heat into electricity was a 10 MW (electric) pilot plant at Barstow, California. The system consists of several principal components: (i) heliostat subsystem, (ii) receiver subsystem, (iii) heat-transport subsystem, and (iv) thermal storage subsystem.

The heliostat subsystem consists of a large array of sun tracking, optically precise mirrors which focus light on the central receiver. The receiver subsystem at the top of the tower has a heat-absorbing surface by which the heat-transport fluid is heated. In the heat-transport subsystem, liquid water (i.e. turbine condensate) under pressure, enters the receiver subsystem, absorbs heat energy, and leaves as superheated steam; typical steam conditions include temperatures around 500°C and a pressure of about 100 atm. The steam is piped to ground-level where it drives a conventional turbine-generator system, and unused heat is rejected to condensor cooling waters. In more advanced systems, the heat-transport fluid might be liquid sodium or a molten mixture of salts at about atmospheric pressure. In these systems, heat would be transferred to water in a heat exchanger located at the bottom of the tower. Another possibility being examined is to use a gas as the heat-transport medium and also as a working fluid in a gas turbine. The purpose of the thermal storage subsystem is to store solar heat energy absorbed in the receiver for use at a later time. The stored energy can be used to produce steam when direct solar radiation is not adequate or available.

3. Environmental Issues

Some hazards, such as potential damage to human eyesight from reflected light, are generic in scope. Others, such as potential occupational exposure to solar thermal process fluids are technology dependent. These issues are explored below.

Air Pollution: Under normal operating conditions, there will be no air pollutants released. In the event of an accident, for example, a fire or rupture of a high pressure valve, by-products from the heat transfer and storage subsystems may be released to the atmosphere. Released pollutants could include nitrogen oxides, sodium monoxide/peroxide, and sodium hydroxide mists or dusts (11, 13).

Water Pollution: Flushing of the heat transfer and storage systems may lead to the planned or accidental release of water pollutants at the plant site. The nature and quantity of these pollutants will vary by system type. Possible pollutants include hydrocarbon oil, corrosion inhibitors (e.g. chromate, borate, nitrate, nitrite, sulphate, arsenate, benzoate salts, triazole, silicate and phosphates), bactericides and glycols. In addition, large amounts of water may be required for steam condensation, boiler make-up and heliostat washing; DOE estimates a demand of 23.6 l/GJ (11).

Land Use: Land use requirements for solar thermal electric plants will vary according to engineering (e.g. system efficiency) and natural (e.g. annual insolation) factors. The U.S. DOE has estimated that a 100 MW (electric) plant constructed in the southwestern United States will require 250 ha/10^6 GJ (11).

Visual Impacts: The visual impact of large solar thermal electricity plants will probably be similar to those of conventional sources. The need for direct sunlight, however, might make them less easy to disguise by landscaping.

Health Risks: Health risks -- public and occupational -- associated with this specific energy system are reflected light and working fluids. One problem with central-station receiver might arise from the danger to human eyesight from light reflected by the heliostat field. Sunlight from a single concentrating heliostat field could quickly cause blindness, and a large-scale generating plant will have a large number of these devices. Most will be sun-tracking, and failure of the control mechanisms could cause the beam to be deflected away from the central receiver. Even in normal conditions, this could pose a health risk to workers at the site, and might present constraints on the use of land within or near the boundaries of such plants. Workers' health could also be adversely affected by failures of the heat transfer and storage subsystems. The nature and size of these hazards will depend on the type of accident and of heat transfer fluids used. Of the latter, important determinates of hazard include temperature, pressure and toxicity of the transfer fluids released (13).

Noise: There are no particular noise problems associated with solar thermal generation, though it shares some of the characteristics of other steam-generating plants.

Thermal Discharge: Like most thermal generating plants, solar thermal units may produce waste heat which must be discharged into local watercourses or into the atmosphere.

Other Impacts: In addition to the direct commitment of lands for the heliostat and receiver subsystems, installation and operation of solar thermal electric plants may produce several other effects related to shading, wind deflection, and soil compaction. When vegetation has to be removed and the soil compacted, erosion may be a problem. Panel shading and wind deflection may produce small micro-climatic changes, including decreased temperature (14-17).

It has been suggested that local climatic effects of large facilities may be substantial, although their significance and size have not yet been evaluated (6, 17, 18).

4. Conclusions

Under normal operating conditions, the most important environmental effects of solar thermal electric power plants probably relate to the large land use needs of these facilities and the subsequent effect on the local biota due to changes in soil structure, microclimate, etc. Two unique hazards to health from these systems may arise from reflected light, and the accidental release of heat transfer fluids. Misguided or accidentally reflected light might cause eyesight injuries. Workers exposed to such potential hazards could, however, be protected by wearing special glasses. Important determinant of risk from heat transfer fluids include temperature, pressure and toxicity of the transfer fluids released (13). These will vary by system type.

V. PHOTOVOLTAIC ENERGY SYSTEMS

1. Introduction

The physical phenomenon responsible for converting sunlight to electricity -- the photovoltaic effect -- was first observed in 1839 by a French physicist, Edmund Becquerel. He noted that a voltage appeared when one of two identical electrodes in a weakly conducting solution was illuminated. In the late 19th Century, the photovoltaic effect was first studied in solids such as selenium when photovoltaic cells were built that converted light to electricity with 1 to 2 per cent efficiencies (19). Over the years, knowledge of the underlying physics of the photovoltaic phenomenon expanded. In the 1920s and 1930s, quantum mechanics provided the theoretical foundation for the present understanding of the photovoltaic effect. A major step forward in solar-cell technology was development in the early 1950s, of a method for producing highly pure crystalline silicon, which was used by Bell Telephone Laboratories in 1954 to produce a silicon photovoltaic cell with 4 per cent efficiency. Bell Laboratories soon produced cells with 11 per cent efficiency, heralding an entirely new era of power-producing cells (20).

In the 1950s, a few schemes for commercial application of silicon photovoltaic cells were tried, mostly in regions geographically isolated from electric utility lines. These schemes were not very successful, but an unexpected boom arose from a different quarter in 1958, when the U.S. Vanguard space satellite was launched. This satellite used a small (< 1 W) array of photovoltaic cells to power its radio. The cells worked so well that space scientists soon realised that they could be an effective power source for many space missions. Photovoltaic energy systems have been an integral part of the U.S. space programme ever since (19, 20).

Photovoltaic cells have also been used in terrestrial applications since the late 1950s, but only since 1975 has the market for earth-based systems outstripped that for space systems. Currently, photovoltaic systems are penetrating, stepwise, three different markets: (i) small applications in electronic equipment and novelty items such as calculators; (ii) small remote non-grid-connected applications such as mountain-top radio repeaters; and (iii) electric power systems comprising small ($< 10\ kW_p$), intermediate ($< 500\ kW_p$) and large ($1\text{-}100\ MW_p$), installations serving single residences, commercial and industrial settings, and central-station generators, respectively (21). The high cost of these systems has been the main hindrance to more extensive use in these applications, but the rising costs of alternative energy supplies along with the decreasing costs of photovoltaic devices are now resulting in rapid expansions into these markets (Table 2.3). (kW_p or MW_p are expressed as "peak" power.)

Table 2.3

WORLD-WIDE MARKET PHOTOVOLTAIC MARKET SHARES BY END-USE CATEGORY (%)

	Year	
End-use category	1981	1982
Remote Stand-Alone	76	52
Central-Station	0	24
Intermediate	6	6
Residential	1	1
Specialty	17	17
	100	100

The photovoltaic's industry has grown in size as new market places appear. Currently, the solar cell industry has an annual growth rate of 75 per cent. Worldwide cell production has gone from a cumulative annual production of about 1 MW_p in 1978 to about 22 MW_p in 1983. Similarly, estimated sales rose from about $85 million in 1980 to between $350 to $400 million in 1983 (22). The industry has grown rapidly within and outside the USA. Of the approximately 40 solar module manufacturers, the USA has 6, Europe has 12, Japan has 14, and approximately 10 companies are producing photovoltaics on a small scale in developing countries. Although the US firms first initiated production on these devices and as recently as 1980 produced 80 per cent of the world's solar cells, production in other countries has advanced rapidly in recent years. In 1983, European companies produced 3 MW_p of solar cells, Japanese 5 MW_p and United States 12 MW_p. Although US firms produced a greater quantity of solar cells in 1983 than any other country, the Japanese led in sales with total revenues of $200 million (20).

2. Technology Review

A typical flat-plate solar cell of present design is about 100 cm^2 in size and contains a flat layer of semiconductor material. When sunlight strikes the cell, electrons are freed in the semiconductor material, and an electric current is generated. The electricity is collected and transported by metallic contacts placed in agrid-like fashion on the surface of the cell. Groups of cells are mounted together on a rigid plate and are electrically interconnected to form photovoltaic modules. A typical module can convert about 11 per cent of the incoming sunlight to electricity and has a generating capacity of 50 W_p. Some modules are designed to concentrate sunlight onto smaller cells, 1 cm^2 or less in area, and can convert 14 per cent or more of the incoming sunlight to electricity. In either case, groups of modules are usually mounted together on a permanently attached frame and are interconnected to form photovoltaic arrays.

Solar cells can be made from a number of materials and formed in a variety of designs; they are classified by material and by type of fabrication process.

Single-crystal silicon is the most frequently used and best understood semiconductor material for photovoltaic cells. In cell fabrication, a Czochralski or a floating-zone process is used to obtain single-crystal silicon from polycrystalline. These processes, particularly the Czochralski, have been the dominant ones in photovoltaics' manufacture since their rediscovery in 1954; in 1982, they were used for more than 60 per cent of all photovoltaic cells produced worldwide. These processes, however, are both expensive and wasteful, therefore new material and process options have been developed; some are now being used commercially, and others are still being studied.

One way to reduce costs is to fabricate the single-crystal cells less expensively by using less-refined silicon and by reducing excessive material losses due to cutting. Edge-defined film-fed growth, ribbon-to-ribbon growth, and dendritic web growth are among the processes for making cheaper silicon cells. Some of these options are now just beginning to be used commercially. In 1982, their share of the market was about 19 per cent; in the near-term, it will continue to grow.

In a continuing effort to reduce system costs further, research has recently been directed towards two different approaches to cell fabrication: thin-films and high-efficiency multi-junction cells. Although still in early stages of development, thin-films offer the potential for making inexpensive cells because they require less material and allow more highly automated production. Amorphous silicon alloys and several polycyrstalline compound semiconductors (e.g. gallium arsenide, copper indium diselenide, cadmium telluride, zinc phosphide) show promise as materials for cost-competitive single-junction thin-film cells and some of them may be used also in multi-junction cells, but so far these materials have shown realtively low efficiency, and in some cases they are unstable. The maximum efficiencies achieved for small, laboratory-scale (1 cm^2), thin-films are 10 to 11 per cent and for larger cells (100 cm^2), 7.5 per cent. These are much higher than those previously achieved but still not as high as those of cells now made by more conventional means, e.g. 11 to 18 per cent for cells made by the Czochralski process (23). Long-term research may yield small-cell efficiencies as high as 18 per cent with use of selected materials -- possibly as high as 25 per cent (22) with some films. Thin-films based on amorphous silicon are now appearing in the market-place. Ultimately, these types of materials and fabrication processes may replace those currently used.

Multi-junction cells are capable of conversion efficiencies as high as 20 to 35 per cent (23). The primary concept under development entails layering different semiconductor materials so that each layer converts a different portion of the solar spectrum into electricity, thus increasing the percentage of photons that can be converted. Initial research has focused on two areas: crystalline multi-junction concentrator cells and amorphous thin-film multi-junction cells. These concepts are still under investigation and have not yet been used commercially.

3. Environmental issues

Health and environmental issues related to photovoltaic energy systems may arise at several stages in the energy cycle: i) extraction, processing, and refining of raw materials; ii) fabrication, installation, operation, and maintenance of devices; and iii) decommissioning of spent devices. Most attention has focused on the second stage because the activities are specific to photovoltaic systems and because they are the major ones involving potential chemical and physical hazards to environment, health and safety. Some hazards, such as the quantity of different pollutants emitted during production of materials needed to make photovoltaic cells are highly technology dependent (24-30). Others, such as electric shock hazards to persons installing or maintaining photovoltaic devices, are more generic (31, 32). The types of environmental impacts associated with the production and use of several different photovoltaic cell types and applications are discussed below.

Air pollution: The fabrication of thin-film photovoltaic cells will require large quantities of gases. Some of these are listed in Table 2.4 by material and process. Many of them are highly toxic (AsH_3, PH_3, SiF_4, B_2H_6), pyrophoric (SiH_4) or flammable (H_2, CH_4). Many gases likely to be used in thin-film cell production are already being used in industry, but the quantities and application modes will differ. The volumes of some gases, for example PH_3, used for large-scale photovoltaic cell fabrication will be larger than those used for all other purposes. Therefore the options for handling these gases and disposing of unreacted portions (the efficiency of different deposition processes range from 10 to 30 per cent) will need careful consideration.

Accidental releases of hazardous air pollutants may result either from leaking from storage, distribution and process systems or from the venting of process and control equipment during abnormal conditions (fire, power failure, etc.). These could present substantial risks to populations living adjacent to these facilities because of the significant quantities of gases used (32).

Water pollution: The deposition processes which have been examined in detail produce no liquid wastes that are obviously hazardous.

Solid wastes: Solid wastes will be produced as by-products of the deposition processes studied (25, 26). These are residuals adhering to the deposition chamber; they must be removed frequently by mechanical or chemical means as a part of the maintenance process. Some of these wastes, for example, cadmium and arsenic compounds, may be hazardous and will require careful handling and disposal in controlled conditions; recycling would be the best solution.

Occupational Health Risks: In photovoltaic cell fabrication plants, the most important hazards to the workforce will probably arise from the large variety of toxic and hazardous gases (see above discussion) and from electrical equipment used in production; risks from mechanical or noise-related hazards appear to be small (25, 26, 28, 30).

Electrical equipment could present spark generation, laser, electric shock, and radio-frequency (RF) hazards to workers if equipment is designed or used improperly. Heating elements and high voltage RF or DC power sources

Table 2.4

AIR CONTAMINANT HEALTH AND SAFETY HAZARDS(25,26)

Gas	Source	Lethal few min. (ppm)	IDLHa level (ppm)	TLVb level (ppm)	Comments
Arsine	GaAS halide CVD	250	6	0.05	Highly poisonous.
Diborane	a-Si glow discharge, CVD, and RS	160	40	0.1	Highly poisonous. Explosion hazard when exposed to heat, flame, or air. It reacts explosively with Cl_2.
Methane	GaAs	Nonlethal	-	-	Fire and explosion hazard when uncontrollably exposed to heat or flame.
Hydrogen	a-Si glow discharge CVD, and CdTe-ED CdTe-CSVT CdTe-CVD Zn_3P_2-MOCVD	Nonlethal	-	-	Fire and explosion hazard when exposed to heat, flame, or oxidizer.
Cadmium telluride	CdTe-CSVT CdTe-CVD	NA	40 mg/m^3	0.05 mg/m^3	Highly toxic. Reacts with acid fumes or moisture to emit toxic cadmium compounds.
Chlorosilanes	a-Si glow discharge	NA	8 000	5	Dangerous; when heated decomposes to toxic fumes (HCl + Si_2); with water and/or steam produces toxic fumes.
Hydrochloric acid	GaAs halide CVD	1 300	100	5 (ceiling)	Noxious and strongly corrosive.
Hydrogen selenide	$CuInSe_2$-sputtering	-	2	0.05	Highly toxic.
Hydrogen sulphide	CdS-sputtering	500	300	10	Highly toxic.
Methane	Zn_3P_2-MOCVD GaAs-MOCVD	Nonlethal	-	None	Fire and explosion.
Nitrogen	a-Si glow discharge, CVD	Nonlethal	-	-	Can react violently with Ti.
Phosphine	Zn_3P_2-MOCVD a-Si CVD	2 000	200	0.3	Highly toxic. Fire and explosion hazard. Emits highly toxic fumes of PO x when heated.
Silane	a-Si glow discharge CVD	Nonlethal	Nonlethal	NA	High fire and explosion risk; it may ignite spontaneously in air and emit highly toxic fumes.
Silicon tetra-fluoride	a-Si glow discharge	50-250	NA	NA	Highly toxic and irritative; when heated or in contact with acids produces highly toxic fumes.
Trimethyl	GaAs	NA	NA	NA	Fire hazard as it may ignite spontaneously in air.
Trimethyl zinc	Zn_3P_2-MOCVD	NA	NA	-	No toxicity data available. Fire and explosion hazard; also when heated emits toxic zinc compounds.
Zinc phosphide	Zn_3P_2-CSVT	NA	NA	-	Highly toxic. Reacts with moisture or acid fumes to emit highly toxic phosphine.

Abbreviations: CSVT, close spaced vapour transport; CVD, chemical vapour deposition; ED, electrodeposition; HWVE, hot wall vacuum evaporation; MOCVD, metal organic chemical vapour deposition; NA, not available; -, not applicable; RS, reactive sputtering; SP, spray pyrolysis; TE, thermal evaporation; TLV, threshold limit value.

a) The concentration immediately dangerous to life or health (IDLH) represents a maximum level for which escape could be made within 30 min without any escape-impairing symptoms or any irreverisble heating effects.

b) The American Conference of Governmental and Industrial Hygienists (ACGIH) threshold limit value (TLV) is the maximum allowed time-weighted average concentrations for a normal 8-hr working day or a 40-hr working week.

will be used in many thin-film deposition processes. Since flammable and explosive gases are also used in these processes, the possibility of electric spark ignition may present an occupational hazard.

The RF plasma systems present two potential hazards: electric shocks from high intensity currents and biological effects from electromagnetic radiation. Non-solid state generators are high voltage equipment which can generate a fatal current if not properly grounded. RF radiation can damage human cells primarily by a thermal mechanism, but it also may present risks to exposed workers even at levels too low to heat the tissues (30).

In photovoltaic cell manufacturing, laser beams may be used to scribe thin layers of cell materials deposited on large substrates to form narrow strips for individual solar cells. Possible exposure of personnel to the beam, or to the electric field of the source, gives rise to occupational safety concerns. The beam, direct or scattered, can be detrimental to the eye. Most lasers have high voltage (10 to 30 kV) DC or RF power supplies, and electric shock from input power or capacitor discharge can be lethal (30).

Thermal hazards refer to risk of fire and of burns due to contact with hot materials or surfaces. Most thin-film deposition methods considered require heating the substrate, and in some cases the feedstock, to temperatures high enough for accidental contact to cause serious burns. In all cases, however, the hot surfaces appear to be well isolated within the reaction chamber, so the likelihood of occupational burns appears to be quite small.

Public Health Risks: As for any type of electrical installation, homeowners or contractors installing, maintaining, or removing roof-top photovoltaic systems may be at risk to electrical shock. Photovoltaic systems, unlike conventional alternatives, begin to generate electricity immediately upon exposure to sunlight. Although grounding or contact with these circuits is unlikely, exposures to photvoltaic generated electricity could produce serious effects. Studies suggest that the voltage generated by 6 average size modules connected in series is sufficient to cause ventricular fibrillation and possible death under normal temperature conditions. In colder weather, the same effect may be produced by fewer modules (31).

Homeowners having roof-top photovoltaic arrays may also face the hazards of fires caused by: (i) short circuits, and (ii) spontaneous combustion due to heat build-up in dead-air spaces. The probability of either has not yet been estimated for photovoltaic systems, but the consequences of such fires have been studied. In the U.S. in 1980, electrical fires were the fifth leading type of fire and the fourth leading cause of fire deaths. Analysis of this issue suggests that the health risk from fire caused by a photovoltaic system is unlikely to present undue societal risk (risk per individual x number of events), but may be significant for dwelling occupants exposed to such risks (31).

Land Use Impacts: Land use impacts of installed photovoltaic systems will differ for decentralised and centralised applications. In small or intermediate, residential or commercial decentralised applications, photovoltaic systems may be mounted on existing roofs (roof-top) or on the ground (ground-mounted) in small enclosed lots. No additional land is required for roof-top applications. Land area requirements for ground-mounted systems will depend on several factors including insolation and system efficiency. Ground-mounted centralisation photovoltaic systems require about 2 ha/MW_p or 3 $km^2/100\ MW_p$ (30). Lands most suitable for large central-station applications lie in areas where annual insolation is high, e.g., southern portions of the United States. Other factors such as regional energy demand and proximity to population centres, however, may influence siting decisions.

Visual Impacts: As with land use impacts, visual impacts will vary by application type. Aesthetically pleasing, small roof-top systems can be incorporated into the architectural designs of new residences and commercial buildings. Retrofit to existing facilities may be more visually intrusive. Visual impacts of large central-station applications will probably be similar to those of conventional power systems.

Noise: There are no noise problems associated with the fabrication or application of photovoltaic energy systems.

4. Environmental Controls

Hazardous Gases: Improper handling and disposal of hazardous gases may adversely affect occupational and public health. Various management options are available to reduce occupational risks: ventilation, automatic valve shutoffs, etc.

Many gases to be used in photovoltaic cell fabrication are now used by existing industries, but the quantities and application modes differ. Under normal conditions, quantities of pollutants generated from photovoltaic plants should be small (Table 2.5). For these gases, the most commonly used air pollution control system is wet scrubbing or thermal incineration. Properly designed multistage scrubbers can be 98 per cent efficient and thermal incineration 100 per cent in removing SiH_4, SiF_4, PH_3, and B_2H_6 emissions from effluent streams. Release of the remaining emissions from scrubbers should result in ground-level concentrations which would not endanger the health of individuals living adjacent to these plants (25, 26, 28, 29).

Accidental release of large quantities of several hazardous gases stored on site could, however, threaten public health. To reduce risks from accidental release of stored gases, only limited quantities of feedstock should be kept on site, gases should be stored in independent well ventilated sheds with appropriate monitoring and leak or fire-prevention devices, and all employees handling gases should have adequate safety training.

Solid wastes: The quantity and quality of solid wastes will vary by production process (Table 2.6). Some solid wastes may be toxic or hazardous and will require controlled disposal.

Occupational Physical hazards: Like the gas handling situation, a variety of engineering (e.g., use of RF radiation shielding) and administrative (e.g., frequent inspection of all wiring) options are available to reduce occupational hazards (25, 26, 28, 29). These options are routinely used throughout industry, and should therefore not present design or engineering barriers to the safe production of photovoltaic energy systems.

Table 2.5

ENVIRONMENTAL CONTROLS FOR TOXIC OR HAZARDOUS ATMOSPHERIC POLLUTANTS FROM PHOTOVOLTAIC CELL MANUFACTURING FACILITIES (25, 26)

Control			Residuals (kg/yr)[a]	
Compound	Source	Technology	Uncontrolled	Controlled
CdTe	CdTe-CSVT	Part. Filter	40	*
	CdTe-CVD	Part. Filter	200	10
Diborane	a-Si glow discharge deposit	KMNO4 or NaOCl	0.5	0
	a-Si reactive sputtering	KMNO4 or NaOCl	0.3	0
	a-Si CVD	KMNO4 or NaOCl	0.9	0
Hydrogen	CdTe-CSVT	Flare stack	8,900	*
	CdTe-CVD	Flare stack	8,900	*
	Zn_3P_2-MOCVD	Flare stack	8,990	*
Hydrogen selenide	$CuInSe_2$ Sput.	NaOh scrubbing	952	10
Hydrogen sulphide	$CuInSe_2$ Sput.	NaOH scrubbing	366	4
Metal vapours	$CuInSe_2$-All	Part. filter	*	*
Methane	Zn_3P_2-MOCVD	Flare stack	960	*
Phosphine	Zn_3P_2-MOCVD	KMnO4 scrubbing	2 772	80
Silane	a-Si glow discharge	KOH scrubber	1 423	7
	a-Si CVD	KOH scrubber	396	3
Silicon tetra-fluoride	a-Si glow discharge		4 620	25
Zn_3P_2	Zn_3P_2-CSVT	Part. filter	*	*

Abbreviations: see Table 2.8

*Insignificant quantity.

a This represents the total cost of control equipment for a 10 MW p facility.

Table 2.6

TOXIC OR HAZARDOUS SOLID WASTES FROM
PHOTOVOLTAIC CELL MANUFACTURING FACILITIES (25, 26)

Compound	Source	Estimated Residuals (per year)
$CuInSe_2$ compounds	$CuInSe_2$-TE	3 430 kg
	$CuInSe_2$ Sput.	2 770 kg
CdS	CdS-TE	2 060 kg
	CdS Sput.	1 660 kg
CdZns	CdZnS-TE	1 820 kg
CdTe	CdTe-CVST	410 kg
	CdTe-CVD	1 980 kg
	CdTe-HWVE	1 970 kg
Si compounds	Si-RS	205 kg
GaAs compounds	GaAs-CVD	9 900 kg
	GaAs-MOCVD	8 830 kg
Zn_3P_2	Zn_3P_2-CSVT	1 140 kg
	Zn_3P_2-MOCVD	653 kg

Abbreviations: see Table 2.4

Public Physical hazards: Safeguards for photovoltaic components such as diodes, diode housing, wiring systems, and mounting frames to prevent electric shock and fire hazards from rooftop photovoltaic arrays have been identified by Underwriters Laboratories and should be incorporated into all photovoltaic devices (32).

5. Conclusions

Photovoltaic energy systems offer considerable potential for electricity generation in the future. The manufacturing stage could affect occupational health, if prudent controls are not applied by the manufacturers of photovoltaic devices because of potential hazards from explosive and toxic gas handling and storage systems. Past experiences with these materials in related industries, show that the application of available controls can satisfactorily reduce potential hazards from these materials. Likewise, RF, laser and electrical hazards to workers can be effectively minimised by application of engineering and administrative controls. These technologies are increasingly being used in various industrial sectors and risks are better and better controlled.

WIND POWER

1. Introduction

Wind Energy Conversion Systems (WECS) capture the energy of air movements caused by the uneven solar warming of large areas of the Earth's surface. Most modern systems, certainly the larger ones, use lift across the face of an aerofoil surface rather than drag forces to turn the blades. This produces much faster rotor speeds, suitable for electricity generation without complex (and energy wasteful) gear trains. Such systems will not operate in low wind speeds (below about Beaufort 3), i.e., 10 km/h) and the high rotor velocity relative to wind speed means that they become liable to damage in high winds. Most systems are therefore designed to cut out at wind speeds above about Beaufort 10 (80 km/h).

2. Siting Requirements

The power output of WECS is determined by rotor size and wind speed. The formula is: $Q = CD^2 V^3 E$ where C is a constant, D is the diameter of the rotor, V is the wind speed and E is the efficiency of energy conversion of the machine (this formula applies only to horizontal axis machines).

Wind speeds therefore dominate the energy output of WECS and this is likely to be a major constraint on siting. The consistency of winds over the year will determine the maximum load factor of the system. Clearly, the windiest sites will be sought after. Normal meteorological measurements are taken at a height of 10 m but wind speeds increase rapidly with altitude. At low levels, the wind is attenuated by surface drag, so that the best sites are likely to be in areas with little ground cover, often on hilltops or shorelines (or possibly offshore in shallow seas), or on arable land, and the WECS themselves will tend to have large-swept areas. Some of the major problems, therefore, are likely to be visual intrusion and conflicts with farming and nature conservation efforts.

An ad hoc government Commission in Sweden is reviewing the introduction of wind power on a very large scale. The scenario is 10 TWh on land and 20 TWh electricity power offshore. The power plants are planned to be placed in clusters and the installed effects in each plant is approximately 2-3 MW.

The use of wind power, on a very large scale, could cause negative impacts on land use. One of the reasons is that the wind mills need to be placed near the sea and therefore may compete with other land uses, i.e. recreation, tourism and leisure activities. Placed offshore, they could cause problems for navigation and fishing.

Sophisticated studies have also been carried out in Denmark in view of the possible installation of large-dimension wind mills on a fairly important scale. The North West cost of Jutland has been selected as a most favourable site.

3. Visual Intrusion

The comparison (Figure 2.1) with 400 kV transmission pylons and four of the present generation of WECS shows the scale of impact, though it must be borne in mind that the windmills would themselves require transmission lines to carry electricity to the grid. In the case of single machines these may not be visible, and at worst would normally consist of an 11 kV (United Kingdom) system carried on wooden poles. Large-scale remotely sited WECS clusters would require transmission corridors in the same way as other power stations. It is difficult to predict how large clusters of machines would be welcomed by planning authorities or local residents, but their combined visual impact would be considerable. In addition, some likely areas for WECS may be sites that are prized for visual qualities and would be opposed by environmental conservation efforts.

The problem may be less that of absolute land taken but rather of site constraints arising from land ownership and conflicting uses, as large areas are needed for clusters of machines.

FIGURE 2.1

VISUAL COMPARISON OF WECs AND CONVENTIONAL STRUCTURES.

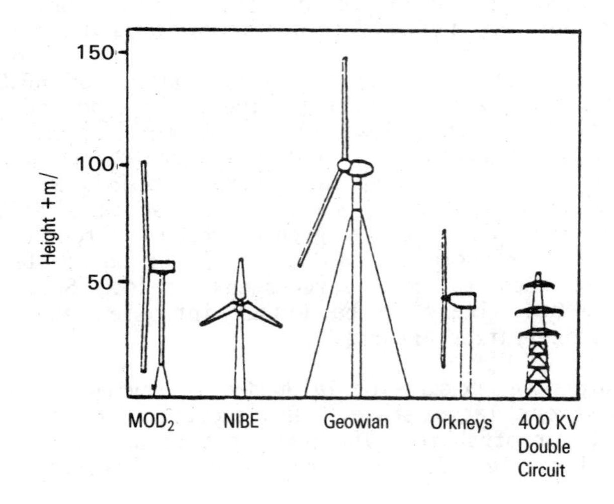

Source: Solar Prospects, the Potential for Renewable Energy, Michael Flood, Windwood House, London, 1983 (7).

4. Noise

A detailed available analysis of noise impacts (33) concluded that noise from rotors could be a substantial constraint on the development of WECS. Certainly, noise problems were encountered by the American MOD 1 unit at Boone. Noise can be produced by the gearbox and generator, but this is well understood and easy to control. The more difficult problem is the noise caused by aerodynamic effects, mainly by fluctuating lift forces on the blade. The most serious effect is on downwind machines (where the rotor is leeward of the tower) and is caused by turbulence in the tower wake. The design of the tower appears to have very little impact on the noise level. Measured noise levels at a MOD 2 site showed that the machine was audible at a distance of 1 400 m upwind and 2 100 m downwind. Noise levels of 56 dB(A) were recorded at a distance of 400 m (34). The "nuisance value" of noise levels is extremely subjective, but it appears that, of the environmental impacts noted, noise is most likely to prove a constraint, especially on larger systems.

The sound levels measured around the Swedish WECS prototypes (Maglarp and Näsudden) are of the same order of magnitude as those measured around the other large units from which data are available, i.e., the MOD-2 and WTS-4 units in the United States. This is easily demonstrated in the A-weighted (expressed in decibels A) sound power level shown in Table 2.7 (35).

Table 2.7

SOUND POWER LEVEL FROM DIFFERENT WECS (35)
(expressed in decibels A)

Unit	Output power	Noise -- dB(A)
Maglarp	3 MW	115 dB
Näsudden	2 MW	116 dB
MOD-2	2.5 MW	118 dB
WTS-4	4.2 MW	117 dB

Neighbours to the Swedish prototypes were polled to determine the noise disturbance during different wind conditions. It was found that high levels of low frequency noise can turn up at a comparatively large distance (35). A thumping sound could sometimes be perceived at a distance of up to two kilometers. There is some evidence that this phenomenon is associated with the sound propagation in the downwind direction. It should be noted, however, that these conclusions are based on a small number of measurements under difficult conditions and on the statements of very few people over a short time period.

Therefore, a comparatively large distance (800 to 2500 m) may be needed between the WECS and residential locations if the same basis for forming a judgment is used as for industrial noise. With suitable amelioration techniques (sound insulation of machinery housing, increased cut-in wind speed, variable rotation speed of turbine), the distance for an upwind unit is reduced to about 300 to 500 m, taking the masking effects of natural wind sound (e.g., wind through trees) into account. An example of a higher cut-in wind speed can be found at the Swedish Maglarp unit (9 m/s or 32.4 Km/h).

5. Radio and TV Interference

Metal rotors can reflect radio waves in a manner that interferes with radio and TV reception. Machines on the line of sight between transmitters and receivers cause amplitude modulations, machines outside the line of sight can cause periodic multi-path signals. All forms of electromagnetic communication can be affected, though more directional systems (e.g. microwave) are only likely to be affected if the WECS is on the line of sight. Apart from the obvious annoyance that this can cause to local residents (avoided by local use of cable systems) there is a fear that WECS could locally interfere with emergency services, air traffic control, aircraft guidance systems, etc. The interference is localised, the precise size of the area affected depending on a wide range of factors. Severe interference can occur at a distance of 2 or 3 km from a MOD 2, though only in a limited range of directions. This has led to unexpected costs in the commissioning of various large scale machines constructed in the USA. Use of the MOD 1 at Boone, North Carolina was restricted during peak viewing hours in early 1980. In Washington, a TV relay station was moved before construction of a WECS cluster at a cost of $160 000 (34). In addition, the installation of a small slave transmitter relieved TV interference at the Swedish Maglarp unit.

6. Other Impacts

The height of WECS towers may be thought to constitute a hazard to aircraft on some sites. In extreme wind conditions some WECS have been known to disintegrate, shedding blades (a 100 kW unit on the Isle of Ushant in Brittany threw one of its 18 m blades 200 meters). It has been calculated that the maximum possible throw distance is 850m (34), and with proper design the likelihood of this happening should be greatly reduced. This might be a constraint in areas with extreme weather patterns, though it should not be a major constraint. Under normal working conditions blades could shed ice -- again throwing it over considerable distances. Ice formation detectors are widely used in the aircraft industry, and have been installed on several large WECS (including MOD 2). It has been suggested that WECS might prove dangerous to low-flying birds or insects, particularly if smaller units with high velocity rotors were collected in clusters. This is unlikely to be a major constraint, except perhaps on known migration routes. In general, the risk that serious accidents will occur with wind power units is considered small.

7. Conclusions

All the impacts of wind conversion appear to be local in scale. This does not mean that environmental considerations might not be a constraint on the development of WECS as local opposition to particular schemes might be considerable. Noise from large scale WECS, especially for downwind machines, can make it difficult to find acceptable sites in areas with scattered housing. Siting in general must determine the compatability of the specific use of the land for wind power generation with other land use interests. In part, the opposition may depend on the use to which units are put. WECS have the capacity to deliver energy in areas remote from main transmission grids, or to link into grids as part of the national electricity supply system. It might be anticipated that local tolerance of the visual and other intrusions would be greater in those applications where the output from the WECS was perceived as being entirely or primarily for local use and if local community draws some economic advantage from it.

BIOMASS

1. Introduction

Biomass, particularly firewood, has for most of man's history been the principal source of domestic heat. Firewood is still used widely in this way, primarily in the less developed countries as well as Canada, Finland and Sweden. However, the biomass technologies now under consideration are intended to provide a range of premium fuels, including liquids for transport uses and gases with a range of calorific values, as well as low-grade heat.

Biomass is a much more complex subject than the other renewable sources, as the term covers such a wide range of potential fuel cycles and therefore of potential environmental impacts. Because of this complexity, it is convenient to treat the production and harvesting of biomass for fuel separately from the conversion and end use.

Many of the environmental questions raised are more normally encountered in the agricultural than in the energy field, and a good deal of analysis from agricultural sources is available. It would not be appropriate here to attempt an analysis of all the environmental impacts that could flow from bringing non-agricultural land into commercial use. Some impacts seem particularly likely to follow from energy uses of land, and these are considered.

2. Biomass Production

Although a wide range of plants and micro-organisms, from algae to forest giants, are being investigated as potential biomass feedstocks, four broad categories can be identified:

-- Organic urban or industrial wastes, principally solid materials and sewage sludge that are currently tipped or incinerated. This source is also known as refuse derived fuel (RDF).

-- Agricultural materials currently treated as wastes, including manure from livestock; straw and other crop residues from agricultural production; bagasse and forestry wastes.

-- Existing uncultivated stands of trees, shrubs, bracken, heather etc.

-- Specially planted energy crops (i.e. sugar cane) either on land brought into production for that purpose, land turned from other agricultural production, or as catchcrops planted on productive land as part of the agricultural cycle.

a) Urban Wastes

The disposal of urban refuse and sewage sludge by conventional means is increasingly expensive and can be environmentally damaging. The conversion of such wastes for energy purposes can, therefore, have a positive environmental effect besides supplying energy. At present, the greater part of the urban waste in most OECD countries is tipped in landfill sites, though a smaller amount is incinerated, for bulk-reduction and hygienical purposes rather than energy-production reasons. Some countries already use a large proportion for heat production and electricity, for example in Denmark about 60 per cent of the waste stream is used.

Urban waste comprises about 65 per cent organic matter, over half of it cardboard and paper. As the quantities of waste increase and the availability of landfill sites is reduced, the economic attractiveness of alternative disposal methods should grow, if only because the burning of waste reduces its volume by 90 or 95 per cent and its weight by 70 to 75 per cent. Refuse can be burned to produce heat and electricity, pyrolysed to produce low to medium calorific value gas, digested to form methane or fermented to produce alcohol. However, refuse is a very variable mix of materials, with varying energy content and a range of potential environmental hazards. Before use in incinerators it requires sorting, drying and shredding.

An energy route that requires the minimum divergence from current disposal practices involves the creation of favourable conditions for the anaerobic decomposition of the rubbish within landfill sites. The production of methane, rather than fatty acids, can be encouraged by site engineering and by the injection of water and nutrients. Gas is then tapped from the site by drilling wells.

The disposal of sewage sludge also presents economic and environmental problems. Untreated sewage is greasy, smelly and can cause severe water pollution, through biochemical oxygen demand and the presence of chemical and biological pathogens. There appears to be no significant epidemiological evidence to suggest that microbiological aerosols such as found in urban waste handling facilities pose a human health hazard (36). Treatment in digesters, as well as producing useful energy, reduces the environmental damage caused by the waste products and helps retain the nutrients, allowing the material to be used as fertilizer.

Another source for refuse derived fuel is industrial waste from wood products and food-processing plants (37). The wood products industry in the USA, for example, currently generates enough wood residues to satisfy its own energy needs at 2.0×10^{18} J/year and still have a surplus. Most food processing wastes, in liquid form, are sent to sewage treatment facilities.

b) Crop Residues and Other Agricultural Wastes

Sources: All agricultural processes produce a range of organic materials with little or no direct commercial value. At present these tend to be disposed of by burning or by leaving them in the soil to degrade. In some cases a proportion at least of these residues is used as bedding or foodstuffs for farm animals, or used in a limited way to provide heat in agricultural buildings or processes. In the forestry industry, only limited amounts of the tree (wood from the bole) are considered merchantable. The foliage, tree tops, branches, stump, and root system are not normally gathered, but are either left on the forest floor or, in the case of full-tree harvesting, by the side of the roadway or landing.

The use of crops for energy purposes allows a much larger proportion of the total plant to be gainfully harvested. This will lead to a drastic reduction in the amounts of crop residues allowed to remain in the soil after harvesting.

Another source of potential biomass energy is manure. Currently, most products are disposed of by land spreading after variable periods of storage as slurries or bedded semi-solids. The use of manure as a fertilizer has a beneficial effect on crop growth and soil conservation and quality. However, animal breeding is often concentrated in specific regions, which may lead to a surplus of manure locally. Under such conditions the manure or liquid slurries are often "dumped" on limited areas, much over the adequate rate and may cause considerable water pollution, especially pollution of groundwaters. The introduction of slurry digesters to produce biogas has, quite apart from the value of the energy produced, a positive environmental effect, as their use conserves nutrients present in wastes and reduces the need for supplementary chemical fertilizers after spreading. The resulting product is high in nutrients and odour emission is reduced, making it particularly suitable as a natural fertilizer. It should be kept in mind that the production of synthetic nitrogen fertilizer is extremely energy intensive, and its overuse often leads to extensive pollution of groundwaters, which are the main source of drinking water. For these reasons the rational use of natural animal fertilizer should be encouraged by agricultural and environmental authorities.

Environmental Impacts -- Although residuals left in the ground seem to have no apparent commercial function, they do of course contribute to maintenance of soil quality. It has been estimated that in the western region of the corn belt of the USA, complete residue removal could cause a 10 per cent decrease in crop yields and soil degradation (37). Crop residues serve a range of functions. Principally, they maintain the organic content and humus of the soil and provide surface protection. This means that they:

-- Control water and wind erosion;

-- Act as a storehouse of nutrients;

-- Stabilise the soil structure and improve its texture;

-- Reduce bulk density;

-- Enhance infiltration and moisture retention;

-- Increase cation exchange capacity;

-- Provide energy for micro-organism activity, an essential factor in soil fertility (38);

There is indeed a risk that significant reduction in the availability of crop residues will adversely affect the quality of the soil.

Soil Erosion -- Soil loss tolerence (the maximum level of soil erosion that will permit a high level of crop productivity to be sustained economically and indefinitely) varies almost from field to field, according to topography, climate, soil type and, not least, according to the use to which the land is put. Soil erosion is already associated with many forms of intensive agriculture, and there is a danger that the removal of crop residues will hasten the destabilisation of the soil. As soils deteriorate, their ability to retain water is also affected, they become more susceptible to erosion and to drought and need increased irrigation.

Nutrient Loss -- The microorganism activity associated with the decay of organic products provides provisional storage for soil nitrogen, and prevents leaching during autumn and winter. If the organic content of the soil is allowed to decline, the ability of the soil to retain nitrogen and water is affected in a number of complex ways. This means that higher rates of irrigation and larger amounts of fertilizer will be applied which will accelerate leaching of nitrate to aquifers and watercourses. Nitrate contamination of aquifers becomes more and more serious, especially in Europe. In conclusion, the use of straw and other residues, on a large scale, for energy purposes may thus have significant impacts.

As opposed to crop residues, logging residues, if they are too abundant and mismanaged, may pose a forest fire hazard, interfere with normal drainage and stream flow patterns, and hinder forest regeneration after logging operations are completed (37). Positive environmental impacts with respect to these problems therefore have been postulated (39). However, the amount of disturbance necessary to remove these scattered residues may also cause erosion problems. Increased biomass removal would, however, decrease the organic and water content of the soil and might affect some ecosystems.

c) Existing Uncultivated Vegetation

Sources: Even in densely populated and agriculturally intensive OECD countries, there may well be scope for developing significant biomass reserves from existing wild plant growth. It has been calculated that the United Kingdom, on the face of it, one of the least promising areas, could in theory meet a large part of its primary energy needs from this source (40). In countries with a long history of commercial timber production, an additional source of otherwise uncommercial biomass could come from degraded forests and from fire- and insect-killed trees. Degraded forests are the result of long periods of selective harvesting of high-grade trees, leaving a residue of lower grade trees. Such plantations are often in need of redevelopment. It has been estimated that over a third of the biomass from timber in Canada could be gathered from these sources.

Environmental Impact -- In principle, this approach would require no presently productive land to be turned to energy use, though it would bring more of the total available land under agriculture or silviculture. There would clearly need to be an infrastructure of roads, conversion plants etc, and both the visual appeal and the ability of the harvested areas to support wildlife would be affected.

The technology used for harvesting energy biomass would be "... closer to that for land-clearing than for logging as the aim is to harvest the maximum possible amount of biomass per unit of land." (41) The environmental consequences of this approach clearly threaten to be severe. Most of the land affected will be more sensitive to disruption than established farm land. The potential impacts on soil quality, nutrient content and erosion are similar to those with crop residue use, though if anything more drastic. Increased biomass utilisation may lead to the acidification of forest soils since the basic cations, once taken up by the trees, will not be returned to the soil in logging residues. Soil disturbance, caused by the intrusion of heavy vehicles and by the removal of tree roots, exposes mineral soil to leaching and encourages erosion by rain and wind. Soil compaction causes a reduction in air and moisure retention and makes seed germination and seedling survival more difficult. Forest clearing exposes the ground to wider extremes of temperature, and leads, of course, to a reduction in organic matter. Some plant species that thrive in the resulting conditions are likely to dominate the area after harvesting, unless careful management is carried out.

The removal of large areas of ground cover could affect both the quality and quantity of available water. Ground cover slows water runoff and aids infiltration into the deeper soil, as well as slowing down the thaw following heavy snow. In extreme cases the disturbance of this cover could cause flooding in low-lying areas. The temperature and nutrient content of watercourses could be altered by harvesting. Fish and aquatic life will be adversely affected by any increased suspended solids, sediments, chemical ions resulting from leaching, and decreased low flows (42).

In addition, terrestrial ecosystems will be impacted. In degraded forests, supported small mammal and bird populations will decrease with removal (42). Large clear cuts also affect game animals because many species use forests for shelter and will not browse or feed in open areas. Habitat reduction may have severe consequences for endangered species and for migratory pathways (39).

Some of the areas affected, especially in more densely populated countries, may be of high recreational use value, and this could be a major cause of conflict. There is unlikely to be any economic conflict, as most of the land will not be in productive use before harvesting. However, once the first harvest has been taken there will be a need for controlled management, including possible use of fertilizers and herbicides, and this may well pre-empt any alternative uses to which the land could be put.

Forest harvesting is known to have great occupational hazards (43). Wood cutting and gathering in North America is a high injury risk job, rivalling underground coal mining. Accidental injury appears 3 to 10 times greater per unit energy from wood than from coal mining (43). However, the wood energy system is free of some chronic health effects (e.g., silicosis and pneumonoconiosis) associated with coal mining (39). Hazards of transport (trucks carrying logs, poles, and lumber) are also significant.

d) Energy Plantations

Sources: There are three principal approaches to the establishment of energy crops. The first involves the use of catch crops, planted between harvests of food crops. The use of industrial crops as one component of an agricultural cycle is a well established practice, as is the planting of fodder crops during the interval between food crops. This latter practice is tending to be abandoned as mixed farming is giving way to more intensive specialised farming, so that an increasing proportion of farmland is remaining unproductive during part of the year. The use of catch crops does not require any additional land to be brought into agricultural use, though it might increase the intensification of agricultural practices in terms of fertilizers, pesticides, irrigation, etc. The second approach involves the fractionation of crops to produce energy materials and food simultaneously.

The third approach involves the creation of specific energy farms, dedicated to the production of fuel materials. Clearly, this requires additional land to be brought into agricultural use, or existing food producing land to be turned to fuel crop production. Estimates of land required range from 20 to 200 x 10^3 km^2 to produce 10^{18} J of biomass energy annually (37). Fuel crops are likely to be economically marginal into the foreseeable future, while food crops will continue to represent the highest added-value use of high-grade agricultural land. It seems likely, therefore, that economic pressures would favour the option of bringing new land under cultivation, probably on a monocrop basis. In some cases this may represent a return to traditional methods; coppicing, for example, is a practice that has been used for centuries, though its use in this context may require a shorter harvesting cycle than is traditional, consisting of two to five year rotations.

Environmental Impact -- The problems that this approach might pose in terms of loss of recreational facilities, the visual intrusion of acres of single crop plantations, depletion of soil quality, the reduction in the gene pool and reduced resistance to disease in intensive monocrop plantations, etc., are well known from large-scale agriculture, though it is difficult to predict the extent to which these problems will apply to fuel cropping until techniques are better established.

Some commentators have gone so far as to query the renewability of biomass (44). In the more intensively farmed areas of the world, fossil fuels have become a significant input into agriculture. In the USA, for example, agriculture is one of the top three energy consuming industries, accounting for three per cent of the national energy budget. Over 90 per cent of the energy used is supplied by petroleum or petrochemicals. It has been estimated that 60 per cent -- 80 per cent of the considerable increase in corn yield per acre since 1945 has been due to energy resource inputs (44).

The danger of enhanced soil erosion if crop residues are turned to fuel production has been mentioned above. However, even in conventional circumstances of intensive agriculture, soil is lost by erosion at a rate that approaches or often exceeds the rate of new topsoil formation. In the USA it has been calculated that the rate of topsoil formation does not exceed about 2.5cm per 100 year (equivalent to 1.5 ton $acre^{-1} year^{-1}$ or 3.4 x 103 kg/ha/yr) (44). At the same time the US department of Agriculture calculated that 97 x 10^6 acres (39 x 10^6 ha) of land in the USA are losing topsoil at rates exceeding the Soil Conservation Service's "acceptable level" of 5 ton $acre^{-1}$ $year^{-1}$ (11.2 x 10^3 kg/ha/yr), which is a very serious problem.

Loss of groundwater resources in intensively farmed areas might be another constraint. (44) Between the end of the 1930s and the end of the 1970s the amount of cropland developed for irrigation in the USA was tripled.

Much of the irrigation water comes from underground aquifers that have only slow natural recharge rates. The economics of aquifer depletion have closely resembled those in petroleum exploitation with little or no incentive to use the resource efficiently. The chief economic constraint on aquifer depletion in recent years has been fuel price rises, which have increased the cost of pumping. Ironically, the development of lower cost renewable energy supplies for pumping water (notably small-scale wind power) may remove this constraint and accelerate the depletion of aquifers.

Most of the environmental impacts of energy farming are comparable to those of food farming. But given the economically marginal nature of farming for fuel, there will be a strong pressure to develop land in the cheapest and most intensive way and impacts are likely to be more drastic.

Occupational hazards of forest harvesting have already been discussed. With respect to energy farms, agricultural and forestry production have occupational risk levels slightly in excess of the national average in the USA (37).

The storage of harvested biomass may also pose environmental problems. Wood fuel piles may become a source of pollution if pile leachate directly enters surface or underground waters (39). Bark decomposition and tannic acid leachate need to be controlled. In addition, odours may emanate from decaying biomass that has been collected and stored for future use (39).

Biomass energy systems, if developd on a large scale, may have the potential to cause changes in local or regional climate conditions (39). Production of biomass, as a result of modifying existing vegetation and soil cover, can alter the properties of the atmosphere (temperature, humidity), the soil (aridity, erosion, albedo, etc.), and hydrology.

Biomass Production Status: Biomass combustion has continued to be the major source of energy in developing countries where it has contributed to disastrous large-scale deforestation. In the USA, it has recently regained popularity. Stocks of biomass from the previously discussed sources were estimated by Morris (37) in Table 2.8. Production from energy farms has not been implemented and depends upon the amount of land to be devoted and species used. Dunwoody et al. (45) predict 100 million acres (40.5 million hectares) of marginal crop land would produce 450 million MT of biomass for approximately 8.9×10^{18} J annually by the year 2000. In Brazil, for instance, sugar cane is cultivated on a large scale for the production of ethanol as automotive fuel. However, even with large available land resources and low labour costs, the resulting fuel has become uneconomic as a result of the recent lower world oil prices.

Table 2.8

BIOMASS GENERATION IN THE USA (37)

Residue	Availability* (10^6 MT/year)	Energy (10^{18} J/year)
Municipal Solid Waste (MSW)	84.0	1.9
Sewage	8.9	0.15
Wood-Products-Industry Residues	130.0	2.6
Large-Feedlot Manures	11.0	0.18
Crop residues	360.0	5.7
Logging Residues:		
-- Above-ground Biomass	77.0	1.6
-- Stumps and Roots	94.0	2.0
Small-feedlot Manures	26.0	0.45

* Availability is defined as the total amount of the resource (dry weight) generated annually in the USA. Due to technical and economic factors, substantially smaller quantities of wastes may actually be available for energy production.

3. Conversion and End Use Technology

It is known that, in contrast to most renewable energy sources, combustion or conversion of biomass materials produces a range of airborne, waterborne and solid emissions, and there is general agreement about their form. Unfortunately, quantification is made extremely difficult by methodological differences between the various analyses. At present, therefore, only relative rankings of the level of emissions will be attempted.

a) Conversion Routes

One of the most important uses for biomass, at least in the medium term, will be its direct combustion for home heating or steam raising. The level of emissions, as is the case with conventional thermal systems, will depend on the type and scale of the technology used. Like coal, biomass can be burned in a range of ways, from highly inefficient open grates to highly efficient fluidised beds. Equally, the level of emissions can be controlled using the same technologies as in coal combustion, for example particulates can be trapped by using cyclones, wet scrubbers, baghouse filters or electrostatic precipitators. Clearly, these techniques will only be economically feasible in larger units. However, the capital costs of biomass systems are already high (biomass boilers cost between three and eight times as much as oil fired) (6), and there will be pressure to avoid additional costs whenever possible. Nevertheless, the centralised industrial boiler is likely to be considerably more efficient, and therefore less environmentally damaging, than individual household units. Biomass conversion technologies and their products are displayed in Table 2.9.

Table 2.9

BIOMASS CONVERSION TECHNOLOGIES -- TECHNOLOGY REVIEW

Process	Products	Premium Fuel
Anaerobic digestion	Methane Carbon Dioxide	Methane
Fermentation/distillation	Ethanol	Ethanol
Chemical reduction and Fractional distillation	Mixed oils	High CV gas or Liquid hydrocarbons
Pyrolysis	Char, oils, tars	Methane, fuel gas Methanol Alcohols
Hydrogenation	Oils	Methane, Liquid hydrocarbons
Direct combustion	Heat	Electricity

b) Air Emissions

Biomass conversion and use give rise to a wide range of air emissions. The type and quantity depends on the conversion process. There is a broad separation between those processes involving biological conversion or wet chemistry and those using combustion or heat treatment, though wet processes carried out at high temperatures will require heating plants, and therefore will share some of the characteristics of combustion systems. Hydrogen sulphide is given off in quantity during anaerobic digestion of animal wastes and to a smaller extent during pyrolysis of crop residues. Ethane may be released to the air during alcohol fermentation. Carbon dioxide is, of course, produced during all combustion as well as fermentation processes (i.e ethanol production), whereas carbon monoxide is produced during combustion processes. Sulphur oxides can be released from fermentation plants or during pyrolysis and direct burning. Nitrogen oxides are associated with most conversion and end use routes, as is particulate emission. In addition, combustion can release a wide range of organic compounds.

Particulates: Biomass, when burned, produces noticeable amounts of ash, part of which becomes entrained with the flue gases. Although particulate emissions can be almost entirely avoided by the use of control technologies, uncontrolled biomass burning is likely to produce larger amounts (perhaps twice the amount) of particulates per unit of energy produced than conventional coal fired systems. Particulate control technology is not likely to be added to very small combustion units, so the principal impact may come from the widespread use of domestic woodburning appliances. Particulates can also be given off, for example, during the preparation of biomass for fermentation, though these are likely to have a larger particle size and therefore be less damaging. Cooper (46) describes wood burning particulates as being almost entirely in the inhalable size range and containing toxic pollutants, carcinogens, co-carcinogens, cilia toxic, mucus coagulating agents, and other respiratory irritants such as phenols, aldehydes, etc.

Carbon Dioxide: In theory, biomass use should not increase the amount of free carbon dioxide in the atmosphere (as in the case of fossil fuels), as the CO_2 released will precisely equal the CO_2 taken up during the growth of the plants. This is partly true under steady state conditions (i.e. when the amount burned equals the amount planted). However, during a period in which large amounts of already-established growth are being converted to energy, free carbon dioxide levels will increase. This also disregards the trapping of carbon in the soil as organic matter, and the fact that extensive use of biomass will probably tend, through various biological and chemical mechanisms, to rapidly diminish such soil organic content.

Sulphur Oxides: Sulphur is not present to any large degree in biomass. The production of sulphur oxides should therefore be considerably lower during biomass combustion than in a coal-fired unit, perhaps by as much as an order of magnitude (see Table 2.10).

Nitrogen Oxides: Nitrogen oxides will be produced in the conversion of biomass, though there is some disagreement as to whether biomass will be more or less polluting in this respect than coal. For some authors, biomass seems likely to produce slightly more NO_x than coal. Biomass has a higher nitrogen content but is often burned at lower temperatures, i.e., lower oxidation rates (37). However, typical Swedish values are 50-150 mg/MJ for biomass compared with 100-350 mg/MJ from coal.

Carbon Monoxide: Release of carbon monoxide from biomass burners is likely to prove higher than from coal, but is highly dependent on the combustion technique.

Emissions from urban refuse incinerators present extra problems because the stream of waste, even after sorting, is likely to be variable and to contain undesirable products. A particular danger is the presence of plastics and chlorinated compounds in general. Burning these can produce a range of highly toxic organic compounds, frequently halogenated, including dioxins, dibensofurans etc. as well as hydrochloric acid in large quantities if PVC is burned. In addition, HCl interferes with electrostatic precipitators (37). Emissions from various conversion routes were estimated in Table 2.10.

Dioxin: Significant concern exists about the presence of dioxin compounds in the fly ash and stack gases of urban refuse incinerators (36). The sources of the dioxin are uncertain; some could be in the waste stream but they are more likely formed during the combustion process (39). Table 2.11 shows levels of organic compounds measured at two refuse incinerators. Greater destruction of the compounds seems to be taking place during co-firing with coal which may be a function of the higher combustion temperatures and exposure time.

Table 2.10

EMISSIONS FROM CONVERSION AND COMBUSTION OF BIOMASS RESOURCES
PER 10^{18} J (37)

	10^4MT						10^6 M^3
	SO_x	NO_x	TSP	HCL	CO	HC	Waste water
Bioconversion Processing(a)	9.9	5	0.33	--	--	--	--
Thermochemical Conversion(b)	37	6.2-49	4.9	4.9	--	--	37-39
Incineration of MSW(c)	12	20	200	6.5	65	N.A.	87
RDF Combustion for Electricity(d)	260-540	65	6.5-65	4.8	29-32	3.6-7.1	--
MSW Shredding	--	--	0.10	--	--	--	--
Industrial Wood Combustion(e)	5.6-9.5	37-64	19-95	--	7.5-390	N.A.	--
Residential Wood Combustion(f)	2.6	6.0	100	--	1 900	N.A.	--

N.A. = Not applicable

a) Biogasification, ethanol fermentation.

b) Pyrolysis.

c) Waterwall incinerator.

d) Co-firing 10 percent RDF with coal.

e) 75 percent efficiency to electricity and heat.

f) 50 percent efficiency to heat.

Table 2.11 presents an estimate of the highest mean annual ground level concentrations by summing the tetradioxin and furan emissions from the mass incinerator and multiplying by an atmospheric dilution factor. This concentration is 4 to 5 orders of magnitude below the expected minimal chronic toxicity and 3 orders of magnitude below the dosage seen to affect lymphocyte activity (36).

Residential Biomass Combustion: In all cases, emissions are likely to be substantially more of a problem if biomass is widely used in household burners. Traditional open fires and unsealed stoves have extremely low efficiencies. They release considerable amounts of particulates, carbon monoxide and hydrocarbons. Sealed stoves are more efficient, though their use of air-control can lead to greater production of smoke and carbon monoxide.

If large quantities of such appliances are used in a particular locality, air quality can deteriorate sharply. An analysis of their use in a small community in the USA showed that the impact of wood burning can be considerably worse than centralised fossil fuel power stations for two reasons. First, residential wood combustion is generally uncontrolled (39). If all 247 houses in the community heat with individual wood appliances for four months, total particulate emissions would equal 16 T. This compares with 12 T produced over the same period by a 1 GW coal-fired power station using up-to-date particulate emission controls. The power station could serve 500 000 people (41). Second, impact will be greater than the power station as residential emissions are released from short stacks with low flue gas velocities. They are therefore more likely to be found in high concentrations at ground level.

Table 2.11

ORGANIC CHLORINE (TOCl) AND DIOXIN EMISSIONS (36)

	Co-firing (a)	Mass Incineration (b)
TOCl, Flue Gas, % Input	0.6%	3.0%
TOCl, Flue, ng/m^3	616	3 200
Dioxins, ng/m^3	No data	505
2,3,7,8,TCDD, ng/m^3	No data	0.41
Tetra CDD*, ng/m^3	No data	6.3
Tetra CDF**, ng/m^3	90.0	

Estimated Highest Mean Annual Ground Level Conc.

$$TCDD/TCDF = 96\ ng/m^3 \times 1.3 \times 10^{-6}\ \text{(atmospheric dilution factor)}$$
$$= 1.25 \times 10^{-13}\ g/m3$$

* Tetrachlorinated Dibenzodioxin
** Tetrachlorinated Dibenzofuran

a) Ames, Iowa coal/RDF cofiring unit burning 10 per cent RDF, 90 per cent coal.

b) Chicago, Illinois waterwall incinerator burning 100 per cent municipal solid wastes.

Wood combustion emissions include a wide variety of organic compounds including acids, aldehydes, phenols, and a range of possible carcinogenic substances, including benzopyrene. It is believed that the solid particles can act as carriers for polycyclic aromatic hydrocarbons. Benzo(a)pyrene (BaP) is often used as an index for estimating risks of cancer per unit exposure. Morris (43) has compared emissions of BaP from different home heating fuels (Table 2.12). Oil heating, depending on the conditions, can be placed equal or somewhat less in terms of BaP cancer risk than wood, and coal heating can vary from slightly less impact than wood to many times greater depending on coal type and furnace or stove operation.

Table 2.12

RESIDENTIAL SPACE HEATING EMISSIONS OF BENZO(a)PYRENE (43)

	μg BaP/GJ
Oil	8.5×10^2
Wood fireplaces	4.3×10^4
Wood stoves	1.3×10^5
Coal	1.6 to 3.1×10^6

The release of such pollutants can clearly create an unfavourable environment within the home as well as externally. The effect on the indoor environment is likely to be exacerbated if the use of inefficient woodburners is combined with high levels of insulation, including weather stripping. Carbon monoxide might be one of the main hazards. CO is normally produced by burning carbonaceous fuels in a limited oxygen supply. However, in the case of a woodburner the provision of adequate air through the bed can lead to CO production, as the carbon dioxide first formed in the combustion is reduced by contact with the glowing embers. The development and wide commercialisation of more efficient and less polluting wood stoves and furnaces is desirable.

c) Solid Wastes

Appreciable quantities of bottom ash and fly ash are produced during biomass combustion. In general, this ash is easier to dispose of than the eqivalent from a coal fired plant, as it contains smaller amounts of toxic trace substances. Ash from urban waste burning, however, can contain a range of toxic substances, including dioxin, though it should be remembered that the unburned waste would in any case have been disposed of by tipping. Nevertheless, some care might be needed in disposing of the ash, as in some cases, tipping ash may cause serious water pollution if it contains heavy metals and toxic substances (39). When spread on agricultural soils, the alkalinity of the ash can improve the growing potential of acidic soils, but may be negative for chalky soils. Slurry wastes from anaerobic fermentation provide a good source of organic fertilizer.

d) Water Pollution

The principal impact on water comes from the biochemical and chemical oxygen demand of the runoff water from both wet processes and combustion processes. In addition, the combustion processes release a certain amount of suspended solids and a range of trace materials, mostly metals and metal compounds. Many of the liquid effluents produced during biomass conversion processes can be used as a chemical feedstock in fertilizer and other industries. Nevertheless, if the waste products are not used in this way they could present a serious environmental problem. It has been calculated that the Brazilian ethanol programme produces 125 million m^3 of stillage in manufacturing 11 million m^3 of ethyl alcohol (the forecast 1985 production level) (47). Two litres of stillage has the same BOD as the sewage produced by one person in a day.

e) Visual Impacts

It is envisaged that much of the medium term use of biomass will be on a relatively small scale. Some types of conversion plant will require most of the components of conventional power stations, including cooling towers.

f) Land and Water Requirements

At the conversion and use end, biomass should not require substantially larger areas of land than conventional energy systems, although larger fuel storage facilities may be required because of the lower energy density of the fuel. In part the land take will depend on the scale of technology introduced, as large numbers of small units are likely to require a larger area of land per unit of energy produced. The water requirements of steam generating electricity sets are likely to be roughly comparable to fossil fuel stations per unit of energy produced, as are those with other conversion processes like methanol conversion. In some cases, for example anaerobic fermentation, the slurry produced can be used as an organic fertilizer. In many cases, with recycling of process water, the consumptive use of water might effectively be small.

g) House Fires

Increased residential wood burning may pose safety risks to the home dweller, due to accidental fires (39, 48). Accidental fires are due mainly to the ignition of flammable materials by hot surfaces or sparks and to chimney fires caused by creosote ignition. Many fires are caused by unsafe installation of stoves and chimneys.

h) Occupational Hazards

Workers in biomass conversion facilities are subjected to different occupational hazards. Fermentation plant workers may be affected by prolonged or accidental exposure to the toxic and corrosive chemicals employed (39). Anaerobic digestion of biomass to methanol produces an unpleasant odour (H_2S) and a potentially hazardous gases (methanol vapours and H_2S are very toxic). In addition, evaporative methanol emissions from storage are highly volatile and could result in explosion and fire if allowed to build up. Waste-processing workers in general have a high accident rate.

i) Public Health Risk

Many toxic substances may be produced and handled during biomass conversion operations. Accidental releases of these materials could pose a significant public health hazard in the vicinity of conversion facilities (37). Gas leaks, explosions, and fires at the facilities or in subsequent fuel handling are possible.

Increased residential heating with wood may entail some risks in terms of injuries associated with wood procurement (cutting, splitting, gathering, etc.) (37). In addition, the use of methanol as automobile fuel will produce toxic emissions. Selective toxicity by methanol can lead to blindness and irreversible neurological damage (39). A U.S. Environmental Protection Agency report lists this as a major concern for biomass energy (49).

4. Environmental Impact Controls

a) Biomass Production

The environmental impacts of biomass production result from harvesting residuals, and new biomass. Amelioration techniques involve prudent agricultural and silvicultural management. Control of soil erosion, for example, can best be achieved by the choice of harvesting system (37). Thinning and selective cutting produce much lower erosion rates than clear cutting systems under all soil, slope and rainfall conditions. If clear cutting is required for silvicultural or economic reasons, patch cutting or strip cutting on the contour will produce much lower erosion rates than full clear cutting under almost all conditions.

The intensity of the use of land for biomass production may be as important as the extent of land use in determining the resulting environmental impacts (37). Land that is exploited for biomass energy production can provide compatible ecosystem services concurrently, depending upon the intensity of ecosystem management. Silvicultural operations using longer rotation periods (for example, on the order of 20 to 50 years) would have significantly lower requirements for agricultural chemicals and other input resources and could support mixed stands of trees. Productivity would be lower, but most other ecosystem values would be less impaired. Partial tree harvesting and coppicing might also be less detrimental.

b) Conversion and End-Use

Amelioration of air pollution at conversion facilities also involves known technologies. Stack scrubbers and electrostatic precipitators can be used. The burning of plastics and subsequent PVC hazards can be avoided with sorting. Dioxin can be destroyed by combustion at higher temperatures.

Residential biomass use is more difficult to regulate. It is known that house fires can be prevented with proper installations, handling and chimney maintenance. Currently, a number of control technologies are being developed that allow production of wood stoves with higher efficiencies and reduced emissions (50). These include: i) catalytic converters, which allow clean combustion at low temperatures; ii) dual-chamber design, which allows for secondary burning of exhaust gases; iii) high-mass stoves, which burn fast and hot, releasing heat over extended periods of time; and (iv) wood furnaces which have forced air and burn quite hot. In the USA, implementation of regulatory strategies to control wood stove emissions is occurring at the local (50), state and federal levels.

5. Conclusions

The impacts of biomass production for energy purposes can be considerable, particularly for soil quality and the resulting ecosystem changes. Prudent harvesting techniques can reduce the effects. Biomass conversion and end-use impacts are in principle similar to fossil fuel burning (i.e., air pollution and ash disposal problems), but some unique hazards depend on the biomass source (e.g., dioxins from solid waste). Alleviation of the public health and environmental risks is possible with known technologies.

GEOTHERMAL ENERGY SYSTEM
(Summarised from References 51-54)

1. Introduction

Geothermal resources are generally defined as accessible heat energy resources in the outer 15 km of the earth's crust. The heat is believed to originate in the earth's interior primarily as a result of the decay of radionuclides.

There are three categories of geothermal resource: hydrothermal, geopressured, and hot dry rock. Hydrothermal resources are underground reservoirs of steam and/or water that contain heat of various grades. Geopressured resources are reservoirs of brime found at great depths that contain energy as heat and pressure, often accompanied by dissolved methane. Hot dry rock is everywhere beneath the earth's crust but is usually too deep for tapping by present technology. In certain areas, deep molten rock has pushed up into the crust and heated solid rock much closer to the surface. These hot dry rock resources may be suitable for development with present technology.

Currently exploited geothermal resources are of the hydrothermal type. The end use and the technology employed depend on whether the resource is vapour-dominated or liquid-dominated. Vapour-dominated (steam) systems are used to generate electricity. Liquid-dominated (superheated water) systems above 130°C are usable for other purposes, such as process or space heating. Reservoirs of lower temperatures (20°-90°C) can be used directly or with heat pumps for space heating or farming (greenhouses etc., particularly in cold countries). Table 2.16 provides a large number of uses of geothermal heat over a temperature range of 20°-200°C.

The geopressured resources are at 3 000-5 000 m below the surface, at pressures between 700-850 km/cm^2. Three forms of energy are potentially available: heat, pressure, and methane. The United States Geological Survey (USGS) estimates that in the USA, some 2.4 million MW-yr are recoverable from hot water, plus about 1 million MW-yr of mechanical energy and about 3.4 million MW-yr from dissolved methane. The quantity of recoverable methane alone is equal to 14 trillion m^3 -- about twice the present proven natural gas reserves in the USA. The geopressured resources programme is currently in a resources definition and experimental drilling stage; the first exploratory wells were only recently drilled in the Texas-Louisiana area.

Hot dry rock is the least developed geothermal resource. If as much as 2 per cent of the energy in hot dry rock alone could be exloited economically, it would supply the entire U.S. non-transportation energy needs at the present rate of consumption. Experimental programmes are in process, but formidable technological problems remain to be solved before hot dry rock systems can be commercialised. The resources are exploited by pumping water down from the surface and recovering superheated water or steam.

Table 2.13

APPROXIMATE TEMPERATURE REQUIREMENTS OF GEOTHERMAL FLUIDS
FOR VARIOUS APPLICATIONS (55)

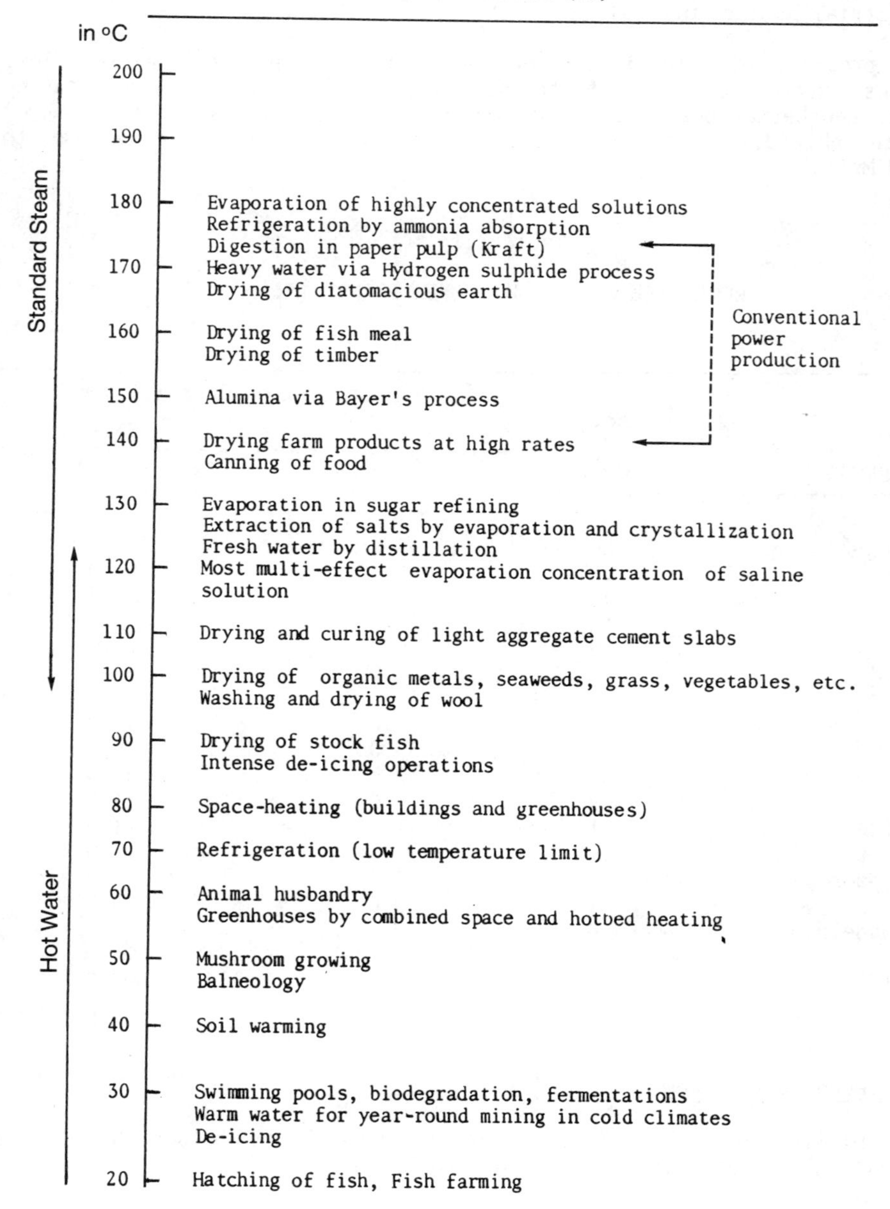
in °C
Standard Steam
Hot Water
200
190
180
Evaporation of highly concentrated solutions
Refrigeration by ammonia absorption
Digestion in paper pulp (Kraft)
170
Heavy water via Hydrogen sulphide process
Drying of diatomacious earth
160
Drying of fish meal
Drying of timber
150
Alumina via Bayer's process
140
Drying farm products at high rates
Canning of food
Conventional
power
production
130
Evaporation in sugar refining
Extraction of salts by evaporation and crystallization
Fresh water by distillation
120
Most multi-effect evaporation concentration of saline solution
110
Drying and curing of light aggregate cement slabs
100
Drying of organic metals, seaweeds, grass, vegetables, etc.
Washing and drying of wool
90
Drying of stock fish
Intense de-icing operations
80
Space-heating (buildings and greenhouses)
70
Refrigeration (low temperature limit)
60
Animal husbandry
Greenhouses by combined space and hotbed heating
50
Mushroom growing
Balneology
40
Soil warming
30
Swimming pools, biodegradation, fermentations
Warm water for year-round mining in cold climates
De-icing
20
Hatching of fish, Fish farming

Some fifty countries around the world are now active or interested in geothermal exploration for electricity generation, for space heating, and for other industrial or agricultural uses.

At present, the total use of geothermal energy in non-electrical applications exceeds that of electrical. As of 1984 the total generating capacity of geothermal power plants in sixteen countries was about 3 825 MW(e) as shown in Table 2.14. Non-electric uses, in contrast, probably amounted to over 6 000 MW(th).

Table 2.14

WORLDWIDE GEOTHERMAL POWER PLANTS (56)

Country	No. Units	Generating Capacity, MW	
		As of June 1984	Expected 1986
United States	30	1 508.0	2 356.0
Philippines	19	781.0	1 496.0
Italy	42	472.0	502.0
Mexico	12	425.0	645.0
Japan	8	215.0	270.0
New Zealand	10	167.0	167.0
El Salvador	3	95.0	95.0
Iceland	5	41.0	41.0
Nicaragua	1	35.0	70.0
Indonesia	3	32.0	147.0
Kenya	2	30.0	45.0
Soviet Union	1	11.0	21.0
China	10	8.0	11.0
Portugal (Azores)	1	3.0	3.0
Turkey	1	0.0	25.0
France (Guadeloupe)	0	0	6.0
Totals	146	3 825.0	5 896.0

2. Technology Review

The ultimate use of geothermal resources depends on whether they are of the hydrothermal, geopressured, or hot dry rock type.

The state of technology for use and conversion of geothermal energy is different for vapour-dominated and liquid-dominated resources. The steam-turbine industry is considered to be in an advanced state of development. Equipment in use throughout the world is producing commercial quantities of electricity from vapour-dominated geothermal sources using technology adapted directly from the fossil-fuel power industry. Furthermore, if the hydrothermal fluid is above 200°C, its well-head pressure can be reduced so that part of the water vaporises (flashes) into steam. Conversion efficiency is about 15 per cent, and, most important, conventional steam turbine technology may be used to produce electricty.

About half the electricity-grade hydrothermal energy in the United States is in this high-temperature range (200°C and above). The other half is in the moderate-temperature range (130-200°C), for which the direct-flash technology may not be economically feasible.

Energy conversion and use processes depend strongly on the nature of the geothermal fluid, principally on whether it is vapour-dominated or liquid-dominated. An important consideration is that certain geopressured fluids contain methane that can be extracted for use, in addition to the heat and pressure energy. The fluid characteristics for hot dry rock resources fall within the range of those for hydrothermal and geopressurised resources, so that the same energy conversion processes are applicable. When geothermal fluids are used for space heating, no turbogenerator is required, so the process is considerably simplified while the environmental impacts are not significantly different. Description of conversion processes will focus on the use of hydrothermal and geopressurised resources for generation of electricity. Although hot dry rock resources are widely distributed geographically, their current development is in the early research stage.

a) Conversion Process for Hydrothermal Resources

The two basic types of geothermal power cycles that are being developed for commercial applications are flashed-steam and binary-fluid. A simple flashed-steam system is depicted in Figure 2.2. Geothermal fluid is withdrawn from a well, and steam is separated (i.e., flashed) from the extracted fluid by pressure reduction. The residual geothermal liquids are disposed of (usually by subsurface injection) while the separated steam is sent to a turbine. Steam exhausted from the turbine is condensed, creating enough water to meet the cooling water needs of the facility. Non-condensing gases are normally ejected from the condenser and if necessary piped to an abatement system. The binary-fluid cycle (Figure 2.3) does not use steam to drive a turbine; instead, down-hole pumps withdraw geothermal fluid from production wells and then the pressurised fluids are sent through a heat exchanger that heats and vaporises a secondary working fluid (e.g., isobutane). The working fluid is subsequently expanded through a turbine, condensed, and reheated for another cycle. Spent geothermal fluids are disposed of by subsurface injection. One advantage of this type of power system is absence of gaseous emissions as long as geothermal fluids are kept at pressures high enough to prevent volatilisation of gases. Moreover, binary-systems are capable of higher conversion efficiencies than flashed-steam facilities and consequently, smaller amounts of geothermal fluids are required per heat unit of electricity generated. On the other hand, binary-fluid facilities must rely on external sources of cooling water because of the lack of steam condensate.

FIGURE 2.2.

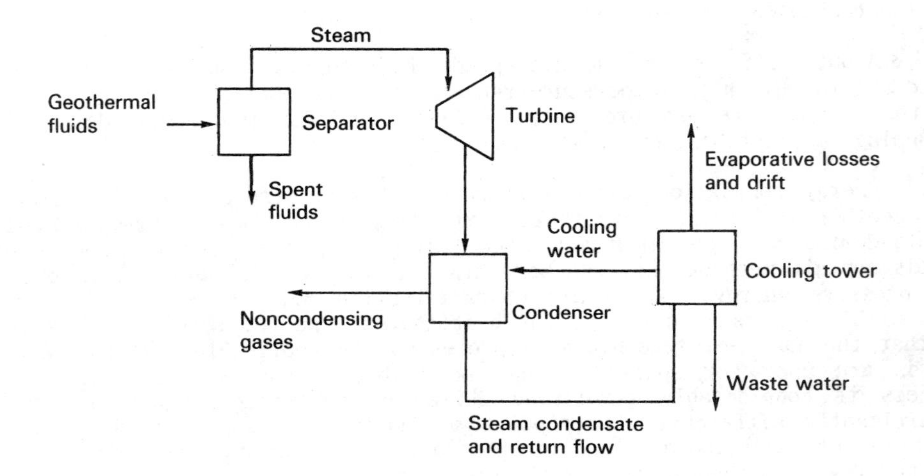

Single-stage, flashed-steam geothermal power cycle. Steam separated from the extracted geothermal fluids drives a turbine-generator to produce electricity. Non-condensing gases ejected from the condenser represent the most important pollutant released from this type of conversion technology (51).

Figure 2.3.

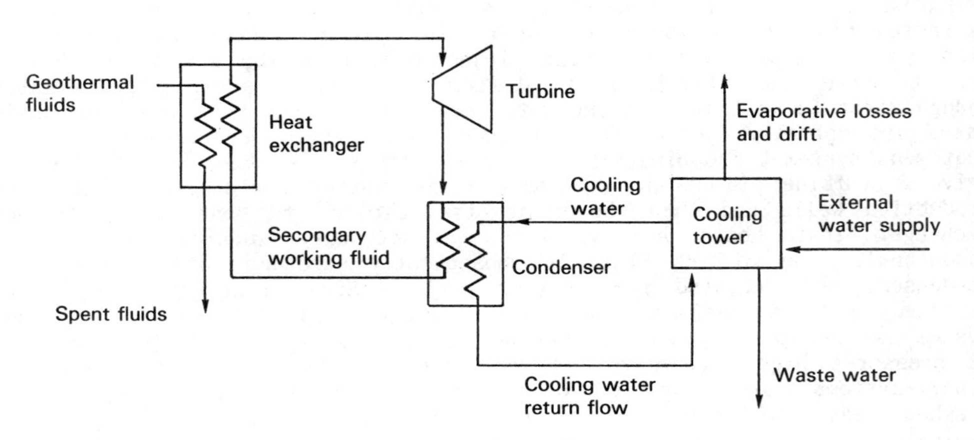

Binary-fluid power cycle. Geothermal fluids are sent through a heat exchanger that vaporises a secondary working fluid, which in turn expands through a turbine-generator to produce electricity. Atmospheric emissions are not expected from this conversion technology as long as the extracted fluids are kept at pressures that prevent the dissolution of gases (51).

Aside from design differences between the two conversion cycles, the primary determinant of resource requirements at a constant heat-rejection temperature is temperature of the geothermal fluid. More precisely, as the temperature of a geothermal resource increases, the efficiency of converting the associated heat energy to electricial energy also increases, thereby reducing the demand for fluid. This relationship is important because the gaseous emissions from flashed-steam power plants are primarily a function of the fluid extraction rate and the concentration of non-condensing gases in the geothermal fluid. Therefore, the lowest gaseous emission rates would be from power plants using high temperature resources containing low concentrations of dissolved gases.

b) Conversion Process for Geopressured Resources

Geopressured resources differ from hydrothermal water-dominated systems principally in that: i) they contain recoverable methane; and ii) the reservoir pressure may be as much as two orders of magnitude higher, on the order of 700 kg/cm^2. These features necessitate additional steps in the energy conversion process. It is believed that the technology can be adapted from existing experience, but no facilities of this type have ever been built.

Methane dissolved in the geothermal fluid is recovered or separated from the fluid in two steps. In the first, the remaining hot water is routed to a high-pressure turbogenerator to produce electricity. In the turbine, the water/methane mixture is expanded to lower pressure. Any remaining methane is then separated, recompressed, and mixed with the high-pressure methane for pipeline sales. The hot water that is left is flashed to steam in the same way as for a normal high-temperature hydrothermal fluid. In the second step, this steam drives another turbogenerator to produce additional electricity, and any residual, unflashed brine goes to waste disposal. Geothermal fluid requirements, solid wastes, emissions, and effluents are insufficiently known at this time.

3. Environmental Issues

There are several potential environmental problems in geothermal energy production. The issues that appear to merit most attention are airborne emissions, solid wastes, brine disposal, chemical or thermal pollution of surface and ground waters, noise, induced seismicity, and subsidence.

Environmental effects are highly site dependent as geothermal reservoirs can have a range of geothermal and geochemical characteristics. It is not possible to talk of 'typical' geothermal systems. Therefore, environmental impacts and techniques for their amelioration, can only effectively be considered at individual sites.

a) Release of Airborne Emissions

Geothermal fluids at depth are complex mixtures of water with dissolved gases and solids. As these fluids are withdrawn from a reservoir and processed to produce electricity in flashed-steam facilities, reduction in temperature and pressure cause volatilisation and subsequent release of various gases that do not condense at atmospheric temperatures and pressures. The major gases in the non-condensing gas phase normally consist of carbon dioxide (at around 90 mole per cent), methane, hydrogen sulphide, ammonia, nitrogen, and hydrogen. Concentrations of these major gases as well as minor gases like benzene, mercury, radon, and boron will vary among wells in the same geothermal resource area and also vary in time. In the following analyses, data on concentrations and emissions of hydrogen sulphide, benzene, mercury and radon -- the most important gases from a health effect standpoint -- are reviewed.

Hydrogen Sulphide -- Hydrogen sulphide is found in nearly all high-temperature geothermal fluids (i.e., 150°C). It is probably formed by one or more of the following mechanisms: reaction of sulphur that is present in reservoir rocks with hot water, magmatic exhalation, or thermal metamorphism of marine sedimentary rocks. Concentrations of this gas sampled from geothermal fluids in the USA range from 0.18-60.7 mg/kg. Table 2.15 presents data on concentrations and emission rates of hydrogen sulphide from several water-dominated resource areas.

Atmospheric releases of hydrogen sulphide constitute the most significant public health issue of geothermal energy production. It is a toxic gas, causing death at concentrations above 1 000 parts per million by volume (ppmv) and eye damage at concentrations as low as 50 ppmv. However, the primary concern is its annoying odour, which can be detected by 20 per cent of the population at a concentration of just 0.002 ppmv. A study of 51 geothermal resource areas showed that at least 29 of these are likely to have one or more power plants that emit enough hydrogen sulphide (without abatement) to cause odour-related problems.

Benzene -- Benzene is associated with the gas phase of fluids derived from geothermal reservoirs of sedimentary origin. Non-methane hydrocarbon gases, including benzene, are thought to evolve from the thermal metamorphism of sediments containing organic matter. Benzene has been identified as a leukemogen. Table 2.16 contains data on concentrations of benzene in non-condensing gases for two water-dominated geothermal resources and two vapour-dominated systems of sedimentary origin.

Table 2.15

CONCENTRATIONS OF HYDROGEN SULPHIDE IN GEOTHERMAL FLUIDS AND UNCONTROLLED EMISSION RATES ESTIMATED FOR HOT-WATER AND VAPOUR-DOMINATED GEOTHERMAL RESERVOIRS IN THE U.S. AND ELSEWHERE (51)

Resource Area	Concentration	Estimated emissions ($g/MW_e.h$)
	In liquids (mg/kg)	
Salton Sea, California	3.2	128(a)
Brawley, California	55.1	2 424
Heber, California	0.18	20
East Mesa, California	0.54	60
Baca, New Mexico	60.7	2 125
Roosevelt Hot Springs, Utah	8	304
Long Valley, California	14	826
Beowawe Hot Springs, Nevada	6	348
Wairakei, New Zealand	--(b)	570
Ahuachapan, El Salvador	48	1 580
Otake, Japan	--(b)	524
Matsukawa, Japan	--(b)	5 050-20 800
Cerro Prieto, Mexico	--(b)	32 000
	In steam (wt %)	
Larderello, Italy	--(b)	14 300
The Geysers, California	24.5	1 850

(a) This emission rate has been recalculated.

(b) The hydrogen sulphide concentration associated with the emission rate was not reported.

Table 2.16

CONCENTRATIONS OF BENZENE IN NONCONDENSING GASES FROM GEOTHERMAL RESERVOIRS OF SEDIMENTARY ORIGIN (51)

Resource Area	Concentration (ppmv)
East Mesa, California(a)	85 -379(b)
Salton Sea, California(a)	100
The Geysers, California(c)	0 -45.5
Larderello, Italy(c)	0.3-38

(a) Water-dominated resource

(b) Concentration was originally reported as a wt per cent in the geothermal fluid

(c) Vapour-dominated resource

Mercury -- Mercury is often present in geothermal waters and gases. It is released from geothermal facilities in liquid and gaseous discharges. Table 2.17 contains data on mercury concentrations for water-dominated systems.

Prolonged exposure to elemental mercury released from geothermal facilities may induce neurological disorders. Moreover, when released into the aquatic environment, mercury is transformed into organo-metallic compounds (methyl-mercury) which is concentrated in food chains (fish in particular). Their effects on humans (Minamata disease) and other living organisms is now well known.

Table 2.17

ELEMENTAL MERCURY IN GEOTHERMAL FLUIDS FROM FOUR WATER-DOMINATED RESOURCE AREAS (51)

Resource Area	Concentration (g/kg of geofluid)
Salton Sea, California	1.8×10^{-6}
East Mesa, California	6.0×10^{-6}
Puna, Hawaii	3.4×10^{-6}
Cerro Prieto, Mexico	2.5×10^{-6}

Radon -- Radon (^{222}Rn), a radioactive gas with a half-life of 3.8 d, is a daughter product of the decay chain of naturally occurring ^{238}U. After radon is formed from the decay of ^{226}Ra in near-surface soils and rocks, it diffuses to the atmosphere at rates that are dependent on ^{226}Ra activity, soil properties, meteorological conditions, and soil moisture. At The Geysers, exhalation rates have been measured that range from 2.6×10^{-6} to 150×10^{-6} pCi/m^2 . s. Similar rates have been measured elsewhere in the world. Radon produced deeper in the earth's crust may never reach the surface because ^{222}Rn can dissolve in groundwater where it decays to its daughter radionuclide ^{218}Po, or the rate of diffusion is so slow with respect to its radioactive decay rate that virtually all of the gas is converted by the time it reaches the near-surface environment.

Others -- CO_2 emissions dominate the gaseous emissions in most geothermal fields, though its overall level per unit of power is usually much less than from fossil fuel plant. CO_2 release has little local impact. The total amount released from a considerable expansion of geothermal acitivity may, however, contribute to CO_2 build-up and global climate change.

Ammonia is another gas released from geothermal wells but it is itself not enviromentally damaging as it diffuses quickly. However, it can react with other chemicals to form environmentally harmful substances, for example ammonium sulphate by reaction with H_2S.

Boron is found as a trace element in geothermal fluids, and may present an environmental problem in terms of injury to crops/vegetation adjacent to geothermal power plants.

b) Release of Liquid Effluents

At many locations, management of spent hydrothermal fluids is one of the most important issues affecting development of geothermal energy.

Water from deep geothermal aquifers contains large amounts (up to 30 per cent) of dissolved solids (sodium chloride, sodium sulphate, potassium chloride, calcium carbonate, etc.). Heavy metals can also be present in solution. The dissolved solids, especially silica, can cause technical problems by damaging or blocking pipes, valves, etc., particularly in flashed steam systems. The salinity of water also means that care needs to be taken during development of a well, to prevent cross-contamination of shallow, fresh water aquifers.

A wide range of environmental impacts can flow from release of high-temperature well products into surface waters. Because the efficiency of geothermal systems is low, the reject liquid will be at relatively high temperatures, and its dumping can produce a temperature rise in the receiving watercourse, while dissolved nitrogen can affect the growth of phytoplankton and weeds. These effects can combine to change the ecological balance.

Elevated mercury and arsenic levels are associated with the release of geothermal water into the Waikato River in New Zealand (57). Fish in the river have been recorded with mercury levels up to 0.8 mg/kg, above the permitted level for human consumption. A separate study (44) concluded that arsenic concentration following the direct, though diluted, release of waste water from a new plant on the Waikato would be "...enviromentally unacceptable." This report further recommended that removal of arsenic by chemical precipitation should be considered only as an interim measure, concluding the reinjection of waste water into the aquifer would provide the only acceptable long-term solution.

c) Land Subsidence

The removal of large quantities of fluid from a geologic formation may result in subsidence, or sinking of land. Such subsidence has been common after withdrawal of water or oil. Evidence of such problems can be seen, for example, at the Wairakei, New Zealand power plant, where waste fluids are disposed of into surface waters. A roughly elliptical dish-shaped depresion covering 65 square km has developed, with a maximum rate of subsidence of about 0.4 m/yr. and a total maximum subsidence so far of over 3 m. Natural subsidence might also be affected by geothermal fluid removal. For example, the Imperial Valley (California) is undergoing substantial natural subsidence, as much as 20 cm during 1972-77. This natural subsidence is, however, regionwide and has not caused damage to land or property.

d) Induced Seismicity

Many hydrothermal reservoirs are in regions that experience frequent natural seismic activity. A continuing question in geothermal energy extraction is whether withdrawal and injection of geothermal fluids may increase the rate of microseismic events. Although one study implicates reservoir cooling, it is believed that the most likely cause of induced seismicity is the injection process. Studies have demonstrated that the frequency of microseismic events can be increased by high-pressure injection of fluids.

It is not known what pressures will be required to maintain the injection of spent hydrothermal fluids over long periods of time at liquid-dominated sites, but field trials indicate that the pressures may be well below those that have been domonstrated to induce microseismicity. The possible lubrication of a major fault and the resultant triggering of a major earth-quake is another unresolved issue; careful site selection to avoid major faults should alleviate this concern.

e) Land and Water Use

Single geothermal wells, once developed, need not occupy large areas of land, though large electricity stations, for example 1 000 MW facilities comprising ten 100 MWe sets, may require considerable areas, (20 to 40 km^2 of land). Individual wells occupy as little as 50 m^2, (58) though an average figure of about 1 km^2 has been mentioned elsewhere. Drilling of new sites will require betwen about 1 000 m^2 and 5 000 m^2 land during construction and drilling (58, 59). Water use depends partly on the temperature of the input water. All geothermal systems are rather inefficient, and therefore will release comparatively large amounts of vapour from the cooling system and reject large quantities of warm water per unit of energy produced. Steam systems will require less cooling water, though they will discharge considerable amounts of water to the atmosphere.

f) Noise

Geothermal noise takes several forms. In the first place there is the noise caused by drilling the wells, at about 65-120 dBA at 15 m. Venting of high-pressure steam from vapour-dominated systems during development of a new borehole, and in some cases continuously during operation of the field, produces noise at about 85-120 dBA at 1.5-3 m. Uncontrolled wells can also produce ground vibrations over an area of about 100 m, as happened with a well at Wairakei, though this can be controlled by drilling relief wells.

Noise emissions were cited as a possible constraint on geothermal developments, at least in part as a result of the problems encountered in the operation of The Geysers field in the USA However, there is reason to believe that these fears are exaggerated. The Geysers field yields dry steam, and is thus a geological rarity. Most potential geothermal sites will be predominantly hot water fields and will not carry out high pressure steam venting, though it is possible that this will be required in the developing technologies based on hot dry rock sources. There are a number of discrete noise sources at The Geysers, though they can roughly be divided into three categories: those associated with steam production, those associated with well drilling, and those related to the generation of electricity. Steam venting is the major noise source, reaching measured levels of 110 dBA at distances of 135 m.

4. Environmental Controls

Some technologies for control of hydrogen sulphide in the geothermal industry were adapted from other industrial processes, such as the removal of sulphur from coal gas. Most are not suitable for geothermal application because they are slow or costly, or because of the chemical form of the sulphur waste product. The systems that react chemically with H_2S such as the Stretford process, seem to be better candidates for geothermal application than systems that depend on physical sorption.

In the Stretford process the off gas is scrubbed with a solution containing sodium carbonate, sodium metavanadate, and anthraquinone disulphonic acid. This process was originally designed to remove H_2S from synthetic fuel gases and many Stretford units are operating throughout the world for this purpose. The steam converter, with a Stretford unit for H_2S abatement, is a promising candidate technology which has been well demonstrated in other applications.

At present, another good candidate for H_2S abatement is a process that utilises copper sulphate to scrub hydrogen sulphide from steam before it reaches the turbine. It is suitable both for vapour-dominated resources and for liquid-dominated resources using flashed systems.

For liquid-dominated resources, the binary conversion system together with total spent-fluid injection offers satisfactory H_2S abatement.

a) Control of Liquid Effluents

The volumes of waste liquids and gases from geothermal plants are large. Even if the plant is temporarily shut down, the wells normally are kept operational, so waste production is continuous.

Only two basic methods of liquid disposal exist: direct discharge (surface disposal) or injection. Both methods have been or are being used at geothermal plants around the world.

Historically, geothermal operators have discharged liquid effluents directly into an available drainage system, but the downstream effects and the increased concern for the environment have brought about more stringent controls for environmental, health, and safety reasons.

For several decades, the oil industry has injected brines into oil reservoirs to enhance oil production and to control land subsidence. Injection into geothermal reservoirs is relatively new and is motivated by a desire to enhance production as well as dispose of geothermal liquids. In France, for instance, re-injection is considered as the normal solution (for various environmental and technical reasons).

Two of the most important differences between the oil field brine-disposal systems and those proposed for geothermal use are the volume and flow rate of fluids to be disposed of. It may be possible to design a well to handle these rates for long periods, but the problems of formation plugging, scaling, and corrosion can be expected to escalate with the volume unless the permeability of the subsurface geologic formation is high.

Treating spent geothermal fluid before disposal would reduce the problems of injection or permit direct release for either secondary use or surface drainage. Simple treatments would be settling and filtration or flocculation. More advanced treatments, such as reverse osmosis, electrodialysis, or ion exchange, are sometimes used on liquid wastes for industries other than geothermal. The solids that are generated by the treatment process can themselves present a major disposal problem. The primary considerations, however, will be the cost (which may be high) and the effect of depleting the producing reservoir by not returning the liquids.

b) Subsidence/Seismicity

Subsidence and induced seismic activity are two potentially interrelated geologic impacts of geothermal development. Subsidence is a direct result of the reservoirs compaction that accompanies the pressure decline caused by fluid withdrawal. Subsidence is generally considered a potential hazard for development of liquid-dominated geothermal resources. The single mitigation measure proposed is injection of the withdrawn fluid. In order not to affect the yield of the producing well, the injection well must be located some distance away. It is not clear whether the injected fluid will always be redistributed along the underlying geologic structure in a way that leaves the surface undisturbed. The land may sink at the withdrawing well and may rise under the injection well.

Induced seismicity may be related to subsidence in that there is some evidence that forced injection of fluids induces microseismic events. Important determining factors appear to be the nature of the subsurface geologic structure (sand or rock), the pressure of injection, and the depth of the injection well. The only mitigating measures are to monitor the seismic activity and to modify the withdrawal and injection rates if increased activity is detected.

c) Noise Reduction

It is possible to use rock mufflers, a technique developed by the operators of The Geysers and these, although costly, can greatly reduce noise emissions, particularly the higher frequencies. Effectiveness of various types of mufflers is shown in Table 2.18. Not all steam vents can be muffled, however, as some wells continue to eject rocks at high velocities during normal operation -- indeed the removal of loose rock is one of the reasons for open venting. Other noise sources are similar to those encountered in traditional energy conversion technologies and can be ameliorated by conventional techniques.

Table 2.18

EFFECTIVENESS OF WELL START-UP NOISE CONTROL METHODS (52)

Control Method	Noise Levels (dBA)
Unmuffled	125 at 15 m
Blooie line expander tube	120 at 15 m
Large metal test muffler	100 at 15 m
Cyclone muffler	90 at 15 m
Rock muffler	72 at 15 m

5. Conclusions

The technology for geothermal energy production is highly site specific, and its success depends almost completely on the characteristics of the geothermal resources. The three primary environmental concerns that could impede geothermal development are treatment of airborne emissions (specifically hydrogen sulphide), spent-brine disposal, and subsidence. The rate of development of geothermal resources depends on technological solutions to these environmental problems. Other barriers to geothermal development include extraction and conversion technology limitations.

HYDROELECTRIC (Refs. 60-61)

1. Introduction

Hydroelectric power generation generally needs the construction of a dam and reservoir, which are responsible for most environmental impacts, both positive and negative. The power output of a particular site depends in equal measure on the difference of water levels (upstream and downstream of the dam) across which generation takes place, and the rate of flow of the river. Slightly different technologies are appropriate in different circumstances. Impulse turbines (e.g., Pelton wheels) require heads of 15-25 meters minimum. Reaction (e.g. propeller) turbines can function at heads of as little as a meter. Large flow/low head schemes require larger dams and bigger turbines per unit of power produced.

Views on the environmental impact of hydro seem to vary. Environment Canada (61) argues that "environmental impacts from carefully designed alternate hydro developments would be minimal and should lend themselves to advance remedial measures." Holdren et al., (2) on the other hand"..... regard the question of big dams versus small ones as still open....", while categorising new large dams as "..... arguably the worst electricity option in terms of damage to ecosystems per unit of electrical power ...". It is also important to note a major distinction between hydropower and fossil fuel or nuclear power plants that generate electricity. In the case of fossil fuel or nuclear power plants, environmental impacts can be attributed exclusively as a direct consequence of energy development. For hydropower, many reservoirs are not constructed solely for hydroelectric generation. If a dam is constructed, its environmental implications will remain very similar, irrespective of whether hydropower is or not a reason for the development. It is not always easy, and often impossible, to ascribe the different environmental implications to specific purposes of water development, i.e., irrigation, power generation, flow regulation, etc. It is within this general context that beneficial and adverse environmental impact of hydroelectric projects are examined in this section.

At present, hydropower is a major source of energy for electricity generation, accounting for almost a quarter of the total electric generation input in the world. Based on an estimate made in the 1980 World Energy Conference, the world's hydroelectric capacity is expected to increase four fold for the next thirty-five years, as many countries have intensive plans for developing their hydropower potentials. Since 55 per cent of the economically feasible hydropower has already been developed in Europe, much of this expected increase will occur in non-OECD countries. Table 2.19 provides historical and estimated hydropower development by country groups from 1976 to 2020.

Table 2.19

HISTORICAL AND PROJECTED HYDROPOWER DEVELOPMENT, 1976-2020 (60)

Potential energy in 10^{18} J

Country groupings	1976	1985	2000	2020
OECD countries	3.78	4.49	5.37	7.80
Developing countries	1.17	1.97	4.49	11.80
Centrally planned countries	0.72	1.20	2.88	8.70
World total	5.67	7.66	12.74	28.30

Since the environmental effects of hydro development are many, and the resulting consequences often extend much further than the project area itself, these can be best discussed by dividing the effects on three categories of subsystems: physical, biological, and human. Table 2.20 provides a summary of the possible environmental implications of hydro development for these three subsystems.

2. Physical Impacts on Environmental Subsystems

Hydropower development projects invariably change river and ecosystem regimes, and thus the real question is not whether such developments will affect the environment, but rather how much change is acceptable to society as a whole, and what countermeasures should be taken to keep the adverse changes to a minimum, at a reasonable economic cost, within that acceptable range.

Construction of any dam imposes an artificial barrier against free flow of water. Water starts accumulating in the newly created reservoir behind the dam, and thus inundates an extensive land area, which can no longer be used for other purposes like agriculture or forestry. The resultant flooding changes the water and ecological regimes of the area.

Although the main purpose of a reservoir is to store water, it starts to store sediments contained in the river water entering the reservoir as soon as the filling begins. Reduction of flow velocity increases sedimentation in the reservoir, which causes the river bed upstream to become higher, leading to further flooding problem. Trapping of silt in dams also causes environmental problems downstream. Before a dam is constructed, large amounts of silt are deposited in the river valley or carried all the way to the delta and the seashore. After construction, these sediments are trapped in the

Table 2.20

POSSIBLE ENVIRONMENTAL IMPLICATIONS OF HYDRO DEVELOPMENT (60)

Physical Impacts

Hydrologic System

Water Quantity
- Level
- Discharge
- Velocity
- Ground water
- Losses

Water Quality
- Sediments
- Nutrients
- Turbidity
- Salinity and alkalinity
- Temperature stratification
- Deoxygenation

Atmospheric System
- Evaporation
- Micro-climate

Land System
- Soil and land structure
- Sismic

Biological Impacts

Aquatic Ecosystem
- Benthos
- Zooplankton
- Phytoplankton
- Fish and aquatic vertebrates
- Plants
- Disease vectors

Terrestrial Ecosystem
- Submerged land and vegetation
- Drawdown zone
- Zone above high water level
- Failure impacts
- Loss of animal habitat
- Food chain repercussions

Human Impacts

Production System
- Agriculture
- Wildlife
- Recreation
- Fishing and hunting
- Transportation

Sociocultural System
- Displacement of people
- Political implications
- Anthropological effects
- Social costs
- Archeological patrimony

reservoir created by the dam. As a result, sediment-free water is flowing downstream of the dam at higher velocities, causing erosion of: (i) the river bed and banks; (ii) the delta or estuary; and (iii) the seashores beyond due to the lack of replenishment of sediment from the river. Loss of silt may affect the productive capacity (agricultural) of the river valleys which used to receive regular deposits of sediments every year; this may be particularly important in developing countries (i.e. Nile Valley). Lack of sediments downstream of the dam may also contribute to significant reduction of plankton and organic carbon which, in turn, may reduce the biomass of impacted aquatic populations in the river and marine areas downstream. Fishing may be greatly affected.

Thermal stratification could be another problem, especially for deep reservoirs. During summer, thermal stratification occurs when warmer epilimmion and cooler hypolimmion waters are separated by a narrow thermocline. The stagnant water layer at the bottom of the lake loses dissolved oxygen due to the decomposition of organic matter. This means that the anaerobic biological population slowly starts to dominate, and they reduce nitrates to nitrites and ammonia. This is followed by sulphate reduction. Such anaerobic activities contribute to the formation of gases like methane, ammonia and hydrogen sulphide which are toxic to aquatic life and affect water quality and thus interfere with water use. If the resulting water of a rather poor quality is discharged through a low-level penstock, it has to travel many kilometers downstream before reoxygenation can restore the water quality. This problem can be observed in many places, e.g. reservoirs of the Tennessee Valley Authority.

The construction activities themselves can have noticeable impacts. If the dams constructed are either rockfill or earthfill, large volumes of rock and earth have to be excavated. In order to reduce transportation costs, quarries and borrow pits are often located near dam sites. Large-scale excavation to obtain such materials may leave unsightly scars on the landscape, and increase soil erosion. Thus, for the construction of the LG 2 Dam of the James Bay project, which required about 150 x 10^6 m^3 of fill, and Wreck Cove Hydroelectric Project in Nova Scotia, Canada, where construction activity was confined to a small area, such environmental problems were encountered. Since dams are mostly constructed in remote areas, often new roads have to be constructed which necessitates removal of vegetation and topsoil and causes erosion.

The presence of a large water body often has climatic implications, especially at the microclimatic level. As a general rule, the greater the surface area of the lake, the farther its influence can be observed. Measurements of all meteorological elements at the standard 2-m height indicate that the influence of man-made lakes decreases exponentially from the shore inland. There is some evidence that man-made lakes in tropical climates decrease convection activity, and thus reduce cloud cover. In temperate regions, during the period prior to freezing, steam fog over the lake and often advective fog over shorelines are common conditions. Precipitation may decrease near the reservoir but humidity is generally increased. Meterological studies carried out near reservoirs indicate that the reservoir has a moderating effect on the climate. Summer temperature around the lake is somewhat decreased and winter temperature may increase noticeably, which is beneficial to touristic and agricultural development.

3. Impacts on Biological Subsystem

Hydropower developments can affect the biological subsystems in many different ways, and the effects could be either beneficial or adverse. Lands flooded due to such developments affect animal life since it can inundate large areas of animal habitat.Likewise, seasonal migration patterns of large mammals, like the reindeer, may be disrupted.

With the filling of the reservoir, a riverine ecosystem gradually transforms to a lacustrine ecosystem. The new reservoir could become an important source of commercial fishery. Fish populations are often very large in new reservoirs during their early years where there is little interest in commercial fishery. Sport-fishing could become important, as in the Lake Diefenbaker in Canada, where such a recreation activity did not exist before. But sooner or later production significantly decreases again, and at the same time, the balance among fish species changes and the number of species diminishes, generally as a result of eutrophication.

The effect of dams can be particularly deleterious on migratory species. Both anadromous (those hatched in freshwater but spending adult lives in sea, i.e. salmon) and catadromous fish species (those hatched in sea but spending adult lives in freshwater, i.e. eel) are drastically affected. Even with special equipment, like the provision of fish ladders, fish migration is considerably restricted. Impoundments can clearly eliminate a number of species. The construction of the Telico Dam on the Little Tennessee River in the United States was delayed for more than two years because it would have eliminated the habitat of an endangered species, snail darter (Percina tanasi), which is protected under the U.S. Federal Endangered Species Act.

The natural vegetation of shorelines (both reservoir and river) are often destroyed or radically changed by hydropower development. Often there is a decrease in the number of plant species close to these embankments. Generally changes in the water levels reduce the diversity of species (flora and fauna). If flow rates are constant, standing crop tends to increase.

Aquatic weeds can increase in abundance as the result of hydropower development, especially in the tropics and semitropics. Unless adequate preventive measures are taken, growth of aquatic weed can be very fast. Weeds create several problems. First, water losses are greatly enhanced by their evapotranspiration process. If irrigation is practiced, more water has to be released to ensure adequate water availability in lower reaches. These two factors often account for a tremendous amount of water loss. They further create problems by interfering with the operation and maintenance of hydroelectric generation and pumping stations, and by competing with fish for space, light and nutrients.

Water development schemes have often enhanced or created favorable ecological environments for parasitic and water-borne diseases such as malaria, schistosomiasis (bilharziasis), dengue and dengue haemorrhagic fever, liver fluke infections and bancroftian filariasis to flourish. Weeds tend to increase the incidence of such diseases by providing a favorable habitat to invertebrates like mosquitoes and aquatic snails, which are vectors and intermediate hosts of disease-causing agents. These diseases are not new; for example, schistosomiasis was known during the Pharaonic times. These diseases, which have considerably increased in recent years, are closely linked to extensive perennial irrigation. Schistosomiasis is currently endemic in over seventy countries, and affects some 200 million people. Over 200 million people are presently infected with malaria in the tropics and subtropics and another 350 million are infected with bancroftian filariasis. Similarly, plant growths around reservoirs provide a suitable habitat for tsetse fly to transmit trypanosomiasis to humans and domestic animals.

In contrast to the diseases mentioned, hydraulic developments tend to reduce the incidence of onchocerciasis. The intermediate host, simulum fly, tends to breed in fast-flowing waters, which are often drowned by the construction of dams. However, adequate measures should be taken to ensure that new breeding places do not develop, especially in the fast-flowing waters near spillways.

It is worth reminding that the diseases mentioned above are, to a large extent, not due to hydropower reservoirs alone, but mostly to the large irrigation schemes which may accompany them in tropical and sub-tropical countries.

4. Impacts on Human Subsystem

Naturally all the effects mentioned earlier have direct or indirect impacts on human beings, but the distinction could be made that such developments at first contribute to environmental and ecological changes, which, in turn affect the human subsystem. Under the present subheading, only those impacts on humans that directly stem from hydro developments will be considered.

Construction of dams and reservoirs often inundate large areas of land (many thousands of square kilometers) Many of the major hydropower projects have contributed to social problems by displacement of the local inhabitants. The number of people that had to be resettled due to construction of dams in different parts of the world is shown in Table 2.21. Resettlement of population due to water development projects in many developing countries has not been a satisfactory experience. Inadequate planning, insufficient budgets, incomplete execution of plans and little appreciation of the problems of technology transfer have all contributed to the failure of plans. The large scale resettlement of population may also prove to be a major constraint on the development of potential hydropower sites in some OECD member countries.

Recreation, tourism and leisure activities may also be influenced by such developments. The importance of these effects is still uncertain, but both positive and negative impacts are expected.

Historic monuments and archeological patrimony have often been flooded by the construction of dams. Since ancient human settlements were often built close to water courses, this is a real problem in many instances in Europe.

Table 2.21

RESETTLEMENT OF PEOPLE DUE TO VARIOUS DAMS (60)

Dams and dates reservoirs filled	Number relocated (approximate)	Countries involved
Aswan, 1968	120 000	Egypt and Sudan
Bhakra, 1963	36 000	India
Brokopondo, 1971	5 000	Surnam
Damojar, 1953 (four projects -- 1959)	93 000	India
Gandha Sugar	52 000	India
Kainji, 1969	42 000-50 000	Nigeria
Kariba, 1963	50 000-57 000	Zambia and Zimbabwe
Keban	30 000	Turkey
Koasou, 1971	75 000	Ivory Coast
Lam Pao	30 000	Thailand
II Projects 1963-1971	130 000	Thailand
Nam Ngum, 1971	3 000	Laos
Nam Pong, 1965	25 000-30 000	Thailand
Nancia, 1967	90 000	Pakistan
Netzahualcoyotl, 1964	3 000	Mexico
Pa Mong (projected)	310 000-480 000	Thailand/Laos
Tarbcia, 1974	86 000	Pakistan
TVA (ca. 20 dams) 1930's present	60 000	U.S.A.
Upper Pampanga, 1973	14 000	Philippines
Volta, 1965	80 000-84 000	Ghana

5. Effects of Dam Failures

When dam failures occur, they can cause disastrous flooding of cities and settlements downstream. Hydroelectric energy is the only renewable source that presents hazards on this scale. The accumulations of water in large schemes may cause local seismic activities, including landslips, which may cause failure of the dam itself. Examples of dam failures have been recorded in all regions of the world, especially in seismic zones, and are well documented in literature.

6. Environmental Controls

As a general rule, the effectiveness of control technologies for adverse environmental effects of hydro projects depends on the characteristics of the particular site.

The erosion of the river bed and banks downstream due to siltation in the reservoir can be reduced by various devices to reduce the velocity and force of the clean water. The deleterious effects of dams on migratory (anadromous and catadromons) fish species can be minimized by building fish ladders and oranising fish breeding. To some extent, it is also possible to compensate for the negative effects on fish production by special measures, such as controlling and balancing food supply and fish populations. Weeds, which are a main source of several environmental problems, can be mechanically removed in shallow waters. For submerged weeds in deep water, biological controls through the introduction of fish, snails, or aquatic grass can be applied. Although chemical herbicides have also been extensively used to control weeds, they themselves pose environmental hazards to aquatic organisms and deteriorate water quality. Their long-term effects on aquatic ecosystems and human health militate for refraining their use. Great progress has been achieved in the conception and building of dams, including anti-seismic techniques, but financial restrictions may not always allow the desirable security margins to be applied. While there are various controls for many environmental effects associated with a hydropower project, long-term substainable controls can only be achieved if ecological and environmental principles are explicitly considered in the overall planning process right from the very beginning.

7. Conclusions

There is no doubt that the majority of hydropower development projects around the world have been beneficial from a socio-economic viewpoint. Equally, however, there is no doubt that many of these development projects have contributed to unanticipated adverse environmental effects, some of which could have been eliminated and others significantly reduced in magnitude by using appropriate planning processes. There appears to be a considerable difference of opinion on criteria and techniques by which successes or failures of hydro projects can be judged. Hence, it is not unusual to find a major water development project hailed as a technological triumph by engineers, welcomed as a success in terms of economic efficiency and regional income distribution by economists, but seriously questioned as to its desirability by environmentalists and sociologists. Lack of public participation during the planning and construction phases further complicates the situation. Planners should plan hydro projects within a broad multi-purpose development concept and give adequate emphasis to their environmental consequences while maintaining full interactions with the public during the planning process. Without these analyses and interactions, an important consideration to judge the success of the water development project, may not be achieved. From a realistic viewpoint, hydropower has been, is and will probably continue to be for a significant time by far the major renewable source of energy. It does not emit into the environment a great deal of emissions and wastes as does the use of fossil (CO_2, acid precipitations) or nuclear fuels (radioactive elements) and although its environmental impact is not negligable at all, it is probably in general less serious than other classical sources of energy. In particular, its impacts are generally only local and not regional (as are acid raidoactive emissions) or global (i.e. CO_2).

OCEAN SYSTEMS

1. Introduction

Oceans have a potential to provide large amounts of recuperable energy. Three different systems are already being exploited or are under development, based on: tidal energy, wave energy and ocean thermal energy. The tidal energy is presently the most advanced approach and has been used in full-scale operations since 1966. Wave energy has also been the subject of considerable efforts of research and various techniques have been developed. Ocean thermal energy conversion (OTEC) systems are not as advanced at the moment, but research is continuing in a number of countries. The potential environmental effects of these three systems are rather varied, and are better known for tidal systems, where longer experience in full-scale operations exists.

2. Tidal Energy

Availability of tidal energy is very site-specific and the number of favourable sites is not unlimited. So far only a few tidal power stations are in operation: the first (1966) and largest one (240 MW) is located on the Rance estuary near Saint Malo, Brittany, France; in the North of the USSR a 400 kW unit has been operating since 1967; in Canada, at Annapolis, a 20 MW plant came into operation in 1984; and China has put a 100 MW plant into service in 1986.

Feasibility studies have been carried out for many sites, some of them of high capacity and power, i.e. the large project of the Bay of Fundy in Canada, south of St. Laurence estuary and the Severn estuary scheme in the United Kingdom, near Bristol, with a projected power of 7000 MW.

In spite of the fact that technological maturity and economic profitability of tidal plants are comparable to those of classical hydropower, the implementation of a number of fairly advanced projects has been delayed, due to the reduced growth rate in electrticity demand, the decreasing oil prices, the high initial investment and the increasing environmental constraints.

Favourable tidal sites are often situated on estuaries or similar types of bays and the environmental impacts of a tidal scheme can thus be compared to those of hydropower on a river, although more complex. Being at the interface between fresh and sea water, they cumulate impacts on both media. Impairment of fish reproduction may be particularly important as such sites are often typical spawning areas as well as crucial pathways for fish migration, both anadromous (which live in sea water and reproduce in fresh water) and catadromous (vice versa). They are also breeding zones for many aquatic organisms (crustacea, shellfish etc.). Navigation and recreation may be impaired for obvious reasons, and pollutants or nutrients from upstream may build-up behind the dam and give rise to specific problems (eutrophication, toxicity, etc.). The substantial modifications in flow patterns may significantly alter the rates of erosion and sedimentation in the bay. Preventive measures may be included in the scheme in order to mitigate the impacts on fish migration, navigation, etc.; they will involve sluices, free channels, and other devices. Although corrective measures may involve some costs and management constraints, compromises can be found which minimise environmental impacts, and still retain good economic and energy characteristics for the project.

3. Wave Power

The concept of using wave energy is relatively old, but it was not until the last decade that small power generators were used and demonstration units experimented in view of large-scale application. Several systems have been studied such as: moveable body, oscillating water column and diaphragm. Such systems, which may be based on-shore or off-shore according to the type, generally need to be associated in large clusters in order to produce sufficient power, although small generators can be used individually as in case of navigational buoys at sea.

Japan, United Kingdom and Norway (62, 63) have devoted significant efforts to the development of wave power techniques. Various systems could now be used commercially, ranging from a few kW to several hundred MW; under favourable conditions and in appropriate geographical locations they could prove competitive in supplying electricity.

Although further experience is still necessary to have a full appreciation of environmental impacts of wave energy systems, the following issues may be raised: action on reproduction of fish and aquatic life; effect on the sedimentation rate on shores and beaches; hazards for navigation (especially off-shore systems). But secondary useful applications could also be found such as damping down the waves in harbour areas, or in zones were wave erosion is a problem. In general, environmental impacts would seem relatively benigh for this source of energy.

4. Ocean Thermal Energy Conversion (OTEC)

This is the difference of temperature between warm surface waters and cold deep waters which is exploited in OTEC systems. A gradient of temperature of 20 degrees (at least) is desirable to get sufficient efficiency, and this gradient can only be found in some zones of the ocean belt between the two tropics, where surface water is very warm and where there is sufficient depth (1 000 metres or more) to find cold water (5°C) on the same vertical.

Even though experiments in producing electricity had already been carried out half a century ago, crucial developments mainly took place during the last decade, in France, Japan and USA. Two basic technologies, open cycle and closed cycle systems, have been developed, the latter being considered probably more promising at the moment. Further research and development are still needed to increase the energy efficiency and the competitiveness of this source of electric power.

Environmental impacts will be very dependent on the site, the scale of operation and the technique used. They may include changes in temperature and salinity of surface waters (especially in open cycles) affecting fish and aquatic life locally, as well as water circulation and streams (which may be very sensitive to slight changes in temperature). Pollution problems could arise from the accidental discharges of working fluids, as well as the use of biocides in normal operation to prevent the proliferation of fixed organisms in the system. The large release of carbon dioxide from deep waters has also been suggested. The hazard of OTEC platforms for navigation, and the risks of pollution (from the ship and the platform) from such an accident cannot be neglected. The environmental impacts of such systems seem, however, relatively moderate, at least for small/medium scale operations.

5. Conclusions

A relatively wide range of environmental problems may arise from ocean energy systems, but the main ones seem to be the impacts on fish and aquatic life, navigation and recreation. The relative lack of experience in some of these systems suggests, however, that detailed studies on potential environmental effects and adequate controls on a site-specific basis, are necessary before the installation of such a system, especially if its scale is relatively important. Although the implementation of ocean energy systerms has been modest so far, due to unfavourable energy and economic conditions, over the past decade (slow-down in the growth of electricity demand, lower oil prices), a promising market may develop in the future, closely linked to favourable geographical conditions, islands in particular, and remote coastal settlements (cities, harbours, tourist resorts) that it could be uneconomic to supply with existing grids. The choice of the system will also depend on geographical conditions. OTEC (Ocean Thermal Energy Conversion) systems will necessarily be situated in warm inter-tropical areas and will produce a constant electricity flux, particularly easy to use. Tidal or wave systems will find more favourable conditions in high latitudes (40/60°), but the electricity production will fluctuate according to tidal cycles and weather conditions, thus likely to need a regulating system. The overall environmental impact of ocean systems seems relatively moderate provided the necessary precautions are taken.

Chapter 3

OVERVIEW ON THE ENVIRONMENTAL IMPACTS OF RENEWABLES

The successful application of any energy system, renewable or otherwise, will depend on the interaction of many technical, economic and environmental factors. In the development of the renewable technologies examined in this report, much attention by governments and researchers has focused on technical and economic considerations; environmental and health related issues are often given only secondary consideration. In a theoretical perspective, damage to environment and health from renewable energy systems, if properly identified as the technology matures from research, to pilot-, and commercial-scale applications, can be mitigated. Unfortunately, past experience suggests that the full impact of energy systems are only recognized after adverse effects have reached signficant levels because insuffficent attention has been give to predicting harmful effects in advance and building effective and affordable remedies into the maturing technology as an integrated system. In recognition of the importance of this principle, this overview attempts to provide focus to the information presented in previous chapters by identifying environmental/health issues which could present serious impediments (i.e., negative attributes), or in contrast significant encouragement (i.e., positive attributes), to the progress of the energy technologies examined. No attempt has been made to compare the relative benefits or costs of any single energy system with any other renewable or conventional alternative because of the large number of theoretical and practical difficulties in making such comparisons: System boundaries; geographic variation; temporal variation; type of damage; degree of knowledge and uncertainty.

Renewable energy sources can in general be considered as more environmentally favourable than most other sources. The approach in this report is constructive; by focusing on impediments, efforts can be initiated to develop effective and affordable remedies which can assure the commercial success and the development of these energy systems, while safeguarding the environment. Benefits can be highlighted to argue for the selected systems, and encourage their promotion.

Impediments to and Benefits of Renewable Energy Systems

The environmental issues considered in this overview are divided into several areas examined throughout this report: material use; land use; water pollution; air pollution; solid waste; noise; visual pollution; ecosystem impacts; and risks to occupational and public health and safety.

1. Material Use

Renewable energy systems (e.g., active solar heating) in general convert relatively dilute energy flows into useful energy. Because of the dilute nature of these energy sources, the relatively low efficiency of the conversion devices, and their decentralised distribution, such energy systems often require larger amounts of materials per unit of useful energy delivered. These material demands impose negative environmental, either due to their production process, toxicity, or disposal. The hazards from such impacts are not new to the public and are not likely to be an important determinant in the success or failure of these systems. They also exist for other energy systems.

2. Land Use

Like material use, large land are also necessary to the production of many renewable energy sources. Particular concern has been raised about the land needs for hydroelectric facilities which can inundate large areas and impact natural resources, ecosystems and human settlements. A number of renewable energy types are also quite site-specific and their siting requirements may create potential conflicts.

3. Water Pollution

Hydroelectric reservoirs, change both the chemical and physical qualities of water including downstream water flows (i.e. eutrophication, deoxygenation, temperature). Brines from geothermal energy systems often contain high salt concentration and toxic materials and pose considerable problems when released to natural waters. Some types of solar systems will require routine flushing of working fluids and the widespread disposal of these fluids containing toxic substances might present problems when significant numbers of these systems are installed. Biomass systems may increase siltation from increased soil erosion, and lead to pesticide and fertilizer runoff from energy crop production.

4. Air Pollution

Although most renewable energy sources are free from air pollutant emissions, a few, namely biomass and geothermal, are responsible for air pollution. Like conventional fuels, production of energy from biomass sources results in a range of air pollutants. These are somewhat similar in nature and quantity to pollutants produced during fossil fuel combustion (CO, CO_2, NO_x, particulates). Toxic pollutants may also be produced (such as dioxins) from refuse-dervied fuels, and polyaromatic hydrocarbons from wood burning. Since these fuels may be often utilized in small decentralized applications, the ability to mitigate or control these releases (specific technologies, high stacks) is less than from large centralized facilities. Geothermal energy systems may also produce large quantities of pollutants; hydrogen sulphide releases are of particular interest. Finally the production of photovoltaic devices may result in the accidental release of toxic gases.

5. Solid Waste

Biomass systems are the only renewable sources which produce solid wastes in signficant quantities, but this is a classical problem.

6. Noise

In wind and geothermal energy systems, significant noise problems may result. This may constrain future technology development, particulary in locations proximate to residential areas.

7. Visual

Any system requiring large land surfaces or high installations may result in visual intrusion. Resistance to such systems will depend on the type of facility, the surrounding landscape, and the attitude of the affected community.

8. Ecosystem

As noted, hydroelectric plants and reservoirs profoundly affect the balance of the properities of the water in impoundments and downstream, as well as the aquatic ecosystems. Dams present signficant barriers to the migration of anandramous and catadramous fish species. These reservoirs can also provide habitat for parasites of importance to man. Alternatively, they can provide suitable locations for commercial fishering and recreational purposes. Biomass systems may also have profound effects on environmental quality: Land clearing, mono-crop production, erosion, desertification, etc.

9. Public Health and Safety

Hydroelectric plants may both provide large benefits and produce negative impacts on public health and safety: flood control vs. catastrophic dam failure; drinking water availability vs. water quality degradation; increased food production from irrigation vs. the development of parasitic illnesses. Other renewable energy systems may present particular hazards linked to air or water pollution, accidents, etc., but also considerable benefits on a health and environmental viewpoint compared to classical energy sources (fossil, nuclear).

10. Occupational Health and Safety

Operation of energy systems and fabrication of any device for the production of different types of energy (renewable; fossil; nuclear) will result in accident-related injuries, deaths or chronic diseases. Some activities, such as wood cutting for biomass, although quite classical present a relatively high risk for workers, approaching that of mines. Some others, such as those linked to solar energy are less known but present probably much lower risks.

Compared to fossil fuel and nuclear energy production and use, renewable energy sources can, in general, be considered to produce less impacts on environment and human health. It is, however, most important that their real or potential impacts can be recorded and assessed as carefully as possible so that all effort can be applied to control and mitigate these impacts and promote full development of best techniques.

REFERENCES

1. IEA Energy Technology Policy, 1985.

2. Environmental Aspects of Renewable Energy Sources, Holdren, Morris and Mintzer, Annual Review of Energy, 1980.

3. Energy Statistics 1981/82, IEA 1984.

4. World Energy Outlook, IEA 1982.

5. Small Scale Hydro-electric Potential for Wales, Professor E. Wilson, University of Salford, 1980.

6. Renewable Sources of Energy, IEA 1983.

7. Solar Prospects, the Potential for Renewable Energy, Michael Flood, Wildwood House, London, 1983.

8. IEA Report: Review of Renewable Sources (unpublished draft 1984).

9. Energy Projections to 2010. Technical Report in Support of the National Energy Policy Plan, 1983, US DOE.

10. Association between Gas Cooking and Respiratory Disease in Children, Melia, R.J.W. et al., British Medical Journal, 1972.

11. Energy Technology Characterizations Handbook, US DOE, March 1983.

12. S. Glasstone, Energy Deskbook, DOE/IR/05114-1, Technical Information Center, Oak Ridge, TN, June 1982.

13. S.W. Kahane, D. Morycz, S. Phinney, J. Hill, M. Yamada, and H. Martin, VII. The Toxicological and Health Implications of Solar Thermal Process Fluids, UCLA 12/1265, University of California, Los Angeles, CA, October 1980.

14. G.R. Hendrey, P.D. Moskowitz, D. Patten, W. Berry, and H.C. Conway, Potential Environmental Problems in Photovoltaic Energy Technology, BNL 51431, Brookhaven National Laboratory, Upton, NY, 1981.

15. D.T. Patten, Report of the Workshop on the Biological Impacts of Solar Energy Conversion, Arizona State University, Temple, AZ, 1978.

16. D.T. Patten, Solar Energy Conversion: An Analysis of Impacts on Desert Ecosystem, Arizona State University, Temple, AZ, 1978.

17. I. Mintzer, Integrated-Assessment Issues Raised by the Environmental Effects of Photovoltaic Energy Systems: A Case Study of Centralized and Decentralized Option, ERG-WP-80-5, University of California, Berkeley, CA, 1980.

18. C. Bhumralker, A.J. Slemmons, and K.C. Nitz, Numerical Study of Local/Regional Atmospheric Changes Caused by a Large Solar Central Receiver Power Plant, DOE/CS/20573-1, U.S. Department of Energy, Washington, D.C., 1980.

19. P. Hersch and K. Zweibel, Basic Photovoltaic Principles and Methods, SERI/SP-290-1448, National Technical Information Service, Springfield, Va, 1981.

20. C. Flavin, Electricity from Sunlight: The Emergence of Photovoltaics, SERI/SP-271-2532, National Technical Information Service, Springfield, VA, 1984.

21. D. Costello and P. Rappaport, "The Technical and Economic Development of Photovoltaics," Annu. Rev. Energy 5, 335-356, 1979.

22. A. Krantz, "Industry Trends," presented at the U.S. Department of Energy National Photovoltaics Program Annual Review, Crystal City, Va., Feb. 1983.

23. Photovoltaic Energy Technology Division, Five Year Research Plan 1984-1988 -- Photovoltaics: Electricity from Sunlight, DOE/CE-0072, National Technical Information Service, Springfield, VA, 1983.

24. P.D. Moskowitz, E.A. Coveney, M.A. Crowther, L.D. Hamilton, S.C. Morris, K.M. Novak, P.E. Perry, S. Rabinowitz, M.D. Rowe, W.A. Sevian, J.E. Smith, and I. Wilenitz, Health and Environmental Effects Document for Photovoltaic Energy Systems -- 1983, BNL 51676, Brookhaven National Laboratory, Upton, NY, September 1983.

25. P.D. Moskowitz, V.M. Fthenakis, and J.C. Lee, Potential Health and Safety Hazards Associated with the Production of Cadmium Telluride, Copper Idium Diselenide, and Zinc Phosphide Photovoltaic Cells, BNL 51832, BrookhavenNational Laboratory, Upton, NY, April 1985.

26. V.M. Fthenakis, J.C. Lee, and P.D. Moskowitz, Amorphous Silicon and Gallium Arsenide Thin-Film Technolgoies for Photovoltaic Cell Production: An Identification of Potential Safety Hazards, BNL 51768, Brookhaven National Laboratory, Upton, NY, October 1983.

27. I. Wilenitz, V.M. Fthenakis, and P.D. Moskowitz, "Costs of Controlling Emissions from the Manufacturing of Silicon using Dendritic Web Photovoltaic Cells," Solar Cells, in press.

28. J.C. Lee and P.D. Moskowitz, Hazard Characterization and Management of Arsine and Gallium Arsenide in Large-Scale Production of Gallium Arsenide Thin-Film Photovoltaic Cells, Brookhaven National Laboratory, Upton, NY, April 1985. Submitted to Renewable Energy Sources.

29. V.M. Fthenakis and P.D. Moskowitz, Characterization of Gas Hazards in the Manufacture of a-Si Photovoltaic Cells, BNL 51854, Brookhaven National Laboratory, Upton, NY, April 1985.

30. V.M. Fthenakis, Hazards from Radio-Frequency and Laser Equipment in the Manufacture of a-Si Photovoltaic Cells, BNL 518153, Brookhaven National Laboratory, Upton, Ny, April 1985.

31. P.D. Moskowitz, E.A. Coveney, S. Rabinowitz, and J.I. Barancik, "Rooftop Photovoltaic Arrays: Electric Shock and Fire Health Hazards," Solar Cells, 9, 327-336, 1983.

32. Underwriters Laboratories, Inc., Safety-Related Requirements for Photovoltaic Modules and Arrays, DOE/JPL/955392-2, National Technical Information Service, Springfield, VA, March 1984.

33. Noise Control Needs in the Developing Energy Technologies, D.N. Keast, US DOE, March 1978.

34. Wind Turbines: Their Effect on the Environment, P.T. Manning (CEGB), IVe Congres Mondial pour la Qualite de l'Air, Paris, May 1983.

35. S. Ljunggren, A Preliminary Assessment of Environmental Noise from Large WECS, Based on Experiences from Swedish Prototypes, The Aeronautical Research Institute of Sweden, FFA TN, 1984-48.

36. J.J. O'Toole, et al., Health and Environmental Effects of Refuse Derived Fuel (RDF) Production and RDF/Coal Co-firing Technologies, Ames Laboratory, U.S. Department of Energy, Ames, IA, October, 1983.

37. G. Morris, Integrated-Assessment Issues Raised by the Environmental Effects of Biomass Energy Systems: A Case Study, Energy and Resources Group Report No. ERG-WP-80-6, University of California, Berkeley, CA, March 1980.

38. Crop Residue Availability for Fuel, J.H. Posselius and B.A. Stout, reprinted in Energy from Biomass, edited by W. Palz, P. Chartier and D.O. Hall, Applied Science Publishers, 1981.

39. U.S. Department of Energy, Energy Technologies and the Environment, Environmental Information Handbook, Report DOE/EP-0026, DOE Office of Environmental Assessments, Washington, D.C., June 1981.

40. Natural Vegetation as a Renewable Energy Resource in Great Britain, G.J. Lawson and T.V. Callaghan, reprinted in Energy from Biomass, edited by W. Palz, P. Chartier and D.O. Hall, Applied Science Publishers, 1981.

41. Energy from Forest Biomass, John Stone Associates for the Corporate Planning Group, Environment Canada, December 1983.

42. U.S. Department of Energy, Environmental Readiness document: Wood Combustion, Report DOE/ERD-0026, Washington, D.C., August 1979.

43. S.C. Morris, Residential Wood Fuel Use: Quantifying Health Effects of Cutting, Transport, and Combustion, presented at the American Public Health Association Meeting in Los Angeles, CA, November 1981.

44. Limitations on the Renewability of Renewable Energy Resources, F.E. Sharples, (Oak Ridge National Laboratory)/Environment International, Vol. 9, 1983.

45. J.E. Dunwoody, T.W. Atkins, and R. Costello, Environmental Impacts of the Utilization of Solar Biomass, Mitelhauser Corporation, Downers Grove, IL, 1980.

46. J.A. Cooper, Environmental Impact of Residential Wood Combustion Emissions and its Implications, Journal of the Air Pollution Control Association, Vol. 30, August, 1980.

47. Energy Crops and the Case of Brazil, S.C. Trindade, reprinted in Energy from Biomass, edited by W. Palz, P. Chartier and D.O. Hall, Applied Science Publishers, 1981.

48. L.D. Hamilton, Comparing the Health Impacts of Different Energy Sources, Keynote Address, Proceedings of a Symposium on the Health Impacts of Different Sources of Energy, International Atomic Energy Agency, IAEA-AM-254/101, Nashville, TN, 22-26 June 1981.

49. J.F. Coates, H.H. Hitchcock, and L. Heinz, Environmental Consequences of Wood and Other Biomass Sources of Energy, Report No. EPA 600/8-82-017, Office of Research and Development, U.S. Environmental Protection Agency, Washington, D.C., April 1982.

50. C.C. Travis, E.L. Etnier, and H.R. Meyer, Health Risks of Residential Wood Heat, Oak Ridge National Laboratory, Oak Ridge, TN, 1985.

51. D.W. Layton, et al., Health and Environmental Effects Document on Geothermal Energy -- 1982 Update, UCLR-53363, Lawrence Livermore National Laboratory, Livermore, CA, November 1983.

52. Environmental Impacts of Renewable Energy Sources and Systems, Draft OECD Document for Group on Energy and Environment, November 1984.

53. D.W. Layton, D.R. Anspaugh, Health Impacts of Geothermal Energy, Proceedings of the Symposium on Health Impacts of Different Sources of Energy held in Nashville, June 1981, Published by International Atomic Energy Agency, Vienna, 1982.

54. L.R. Anspaugh, et al., Human Health Implications of Geothermal Energy, UCRL- 83382, Lawrence Livermore Laboratory, Livermore, CA, August 1979.

55. B. Linal, Geothermal Energy, Review of Research and Development, Earth Science, No. 12, UNESCO, Paris, 1975.

56. R. DiPippo, Worldwide Geothermal Power Development: 1984 Overview and Update, Proceedings of the Eighth Annual Geothermal Conference and Workshop, Electric Power Research Institute, Palo Alto, CA, 1984.

57. Release of Arsenic from Geothermal Sources, J. Aggett and A.C. Aspell, University of Auckland, 1978.

58. The Environmental Impacts of the Renewable Energy Sources, ETSU, United Kingdom, Revised March 1979.

59. The Environmental Impacts of Production and Use of Energy, Vol.III, Renewable Sources of Energy, 1980, UNEP.

60. A.K. Biswas, Hydroelectric Energy, Renewable Sources of Energy and the Environment (Chapter 6).

61. The Energy Brief (Comments on Alternative Energy Sources), Corporate Planning Group, Department of Environment, Canada, June 1982.

62. N. Ambli et al., The Kraerner Multiresonant OWC, Proceedings of the Second International Symposium on Wave Energy Utilization, Ttondheim, Norwav, 1982.

63. TAPCHAN Wavepower, Publication of Norwave A.S. Corporation, Oslo, Norway.

64. T. Anderson, E. Bylund, R. Carlsson, A. Peterson, H. Sjors, and N. Sundberg, River Hydropower Environment -- Environmental Effects of Hydropower Development (Sweden), 1977.

65 Environmental Effects of Electricity Generation (COMPASS), OECD Paris, 1985.

WHERE TO OBTAIN OECD PUBLICATIONS
OÙ OBTENIR LES PUBLICATIONS DE L'OCDE

ARGENTINA - ARGENTINE
Carlos Hirsch S.R.L.,
Florida 165, 4° Piso,
(Galeria Guemes) 1333 Buenos Aires
Tel. 33.1787.2391 y 30.7122

AUSTRALIA - AUSTRALIE
D.A. Book (Aust.) Pty. Ltd.
11-13 Station Street (P.O. Box 163)
Mitcham, Vic. 3132 Tel. (03) 873 4411

AUSTRIA - AUTRICHE
OECD Publications and Information Centre,
4 Simrockstrasse,
5300 Bonn (Germany) Tel. (0228) 21.60.45
Gerold & Co., Graben 31, Wien 1 Tel. 52.22.35

BELGIUM - BELGIQUE
Jean de Lannoy,
Avenue du Roi 202
B-1060 Bruxelles Tel. (02) 538.51.69

CANADA
Renouf Publishing Company Ltd/
Éditions Renouf Ltée,
1294 Algoma Road, Ottawa, Ont. K1B 3W8
Tel: (613) 741-4333
Toll Free/Sans Frais:
Ontario, Quebec, Maritimes:
1-800-267-1805
Western Canada, Newfoundland:
1-800-267-1826
Stores/Magasins:
61 rue Sparks St., Ottawa, Ont. K1P 5A6
Tel: (613) 238-8985
211 rue Yonge St., Toronto, Ont. M5B 1M4
Tel: (416) 363-3171
Federal Publications Inc.,
301-303 King St. W.,
Toronto, Ont. M5V 1J5
Tel. (416)581-1552
Les Éditions la Liberté inc.,
3020 Chemin Sainte-Foy,
Sainte-Foy, P.Q. G1X 3V6,
Tel. (418)658-3763

DENMARK - DANEMARK
Munksgaard Export and Subscription Service
35, Nørre Søgade, DK-1370 København K
Tel. +45.1.12.85.70

FINLAND - FINLANDE
Akateeminen Kirjakauppa,
Keskuskatu 1, 00100 Helsinki 10 Tel. 0.12141

FRANCE
OCDE/OECD
Mail Orders/Commandes par correspondance :
2, rue André-Pascal,
75775 Paris Cedex 16
Tel. (1) 45.24.82.00
Bookshop/Librairie : 33, rue Octave-Feuillet
75016 Paris
Tel. (1) 45.24.81.67 or/ou (1) 45.24.81.81
Librairie de l'Université,
12*a*, rue Nazareth,
13602 Aix-en-Provence Tel. 42.26.18.08

GERMANY - ALLEMAGNE
OECD Publications and Information Centre,
4 Simrockstrasse,
5300 Bonn Tel. (0228) 21.60.45

GREECE - GRÈCE
Librairie Kauffmann,
28, rue du Stade, 105 64 Athens Tel. 322.21.60

HONG KONG
Government Information Services,
Publications (Sales) Office,
Information Services Department
No. 1, Battery Path, Central

ICELAND - ISLANDE
Snæbjörn Jónsson & Co., h.f.,
Hafnarstræti 4 & 9,
P.O.B. 1131 – Reykjavik
Tel. 13133/14281/11936

INDIA - INDE
Oxford Book and Stationery Co.,
Scindia House, New Delhi 110001
Tel. 331.5896/5308
17 Park St., Calcutta 700016 Tel. 240832

INDONESIA - INDONÉSIE
Pdii-Lipi, P.O. Box 3065/JKT.Jakarta
Tel. 583467

IRELAND - IRLANDE
TDC Publishers - Library Suppliers,
12 North Frederick Street, Dublin 1
Tel. 744835-749677

ITALY - ITALIE
Libreria Commissionaria Sansoni,
Via Lamarmora 45, 50121 Firenze
Tel. 579751/584468
Via Bartolini 29, 20155 Milano Tel. 365083
La diffusione delle pubblicazioni OCSE viene assicurata dalle principali librerie ed anche da :
Editrice e Libreria Herder,
Piazza Montecitorio 120, 00186 Roma
Tel. 6794628
Libreria Hœpli,
Via Hœpli 5, 20121 Milano Tel. 865446
Libreria Scientifica
Dott. Lucio de Biasio "Aeiou"
Via Meravigli 16, 20123 Milano Tel. 807679

JAPAN - JAPON
OECD Publications and Information Centre,
Landic Akasaka Bldg., 2-3-4 Akasaka,
Minato-ku, Tokyo 107 Tel. 586.2016

KOREA - CORÉE
Kyobo Book Centre Co. Ltd.
P.O.Box: Kwang Hwa Moon 1658,
Seoul Tel. (REP) 730.78.91

LEBANON - LIBAN
Documenta Scientifica/Redico,
Edison Building, Bliss St.,
P.O.B. 5641, Beirut Tel. 354429-344425

MALAYSIA/SINGAPORE - MALAISIE/SINGAPOUR
University of Malaya Co-operative Bookshop Ltd.,
7 Lrg 51A/227A, Petaling Jaya
Malaysia Tel. 7565000/7565425
Information Publications Pte Ltd
Pei-Fu Industrial Building,
24 New Industrial Road No. 02-06
Singapore 1953 Tel. 2831786, 2831798

NETHERLANDS - PAYS-BAS
SDU Uitgeverij
Christoffel Plantijnstraat 2
Postbus 20014
2500 EA's-Gravenhage Tel. 070-789911
Voor bestellingen: Tel. 070-789880

NEW ZEALAND - NOUVELLE-ZÉLANDE
Government Printing Office Bookshops:
Auckland: Retail Bookshop, 25 Rutland Stseet,
Mail Orders, 85 Beach Road
Private Bag C.P.O.
Hamilton: Retail: Ward Street,
Mail Orders, P.O. Box 857
Wellington: Retail, Mulgrave Street, (Head Office)
Cubacade World Trade Centre,
Mail Orders, Private Bag
Christchurch: Retail, 159 Hereford Street,
Mail Orders, Private Bag
Dunedin: Retail, Princes Street,
Mail Orders, P.O. Box 1104

NORWAY - NORVÈGE
Narvesen Info Center – NIC,
Bertrand Narvesens vei 2,
P.O.B. 6125 Etterstad, 0602 Oslo 6
Tel. (02) 67.83.10, (02) 68.40.20

PAKISTAN
Mirza Book Agency
65 Shahrah Quaid-E-Azam, Lahore 3 Tel. 66839

PHILIPPINES
I.J. Sagun Enterprises, Inc.
P.O. Box 4322 CPO Manila
Tel. 695-1946, 922-9495

PORTUGAL
Livraria Portugal,
Rua do Carmo 70-74,
1117 Lisboa Codex Tel. 360582/3

SINGAPORE/MALAYSIA - SINGAPOUR/MALAISIE
See "Malaysia/Singapor". Voir «Malaisie/Singapour»

SPAIN - ESPAGNE
Mundi-Prensa Libros, S.A.,
Castelló 37, Apartado 1223, Madrid-28001
Tel. 431.33.99
Libreria Bosch, Ronda Universidad 11,
Barcelona 7 Tel. 317.53.08/317.53.58

SWEDEN - SUÈDE
AB CE Fritzes Kungl. Hovbokhandel,
Box 16356, S 103 27 STH,
Regeringsgatan 12,
DS Stockholm Tel. (08) 23.89.00
Subscription Agency/Abonnements:
Wennergren-Williams AB,
Box 30004, S104 25 Stockholm Tel. (08)54.12.00

SWITZERLAND - SUISSE
OECD Publications and Information Centre,
4 Simrockstrasse,
5300 Bonn (Germany) Tel. (0228) 21.60.45
Librairie Payot,
6 rue Grenus, 1211 Genève 11
Tel. (022) 31.89.50
United Nations Bookshop/Librairie des Nations-Unies
Palais des Nations,
1211 – Geneva 10
Tel. 022-34-60-11 (ext. 48 72)

TAIWAN - FORMOSE
Good Faith Worldwide Int'l Co., Ltd.
9th floor, No. 118, Sec.2
Chung Hsiao E. Road
Taipei Tel. 391.7396/391.7397

THAILAND - THAILANDE
Suksit Siam Co., Ltd., 1715 Rama IV Rd.,
Samyam Bangkok 5 Tel. 2511630
INDEX Book Promotion & Service Ltd.
59/6 Soi Lang Suan, Ploenchit Road
Patjumamwan, Bangkok 10500
Tel. 250-1919, 252-1066

TURKEY - TURQUIE
Kültur Yayinlari Is-Türk Ltd. Sti.
Atatürk Bulvari No: 191/Kat. 21
Kavaklidere/Ankara Tel. 25.07.60
Dolmabahce Cad. No: 29
Besiktas/Istanbul Tel. 160.71.88

UNITED KINGDOM - ROYAUME-UNI
H.M. Stationery Office,
Postal orders only: (01)211-5656
P.O.B. 276, London SW8 5DT
Telephone orders: (01) 622.3316, or
Personal callers:
49 High Holborn, London WC1V 6HB
Branches at: Belfast, Birmingham,
Bristol, Edinburgh, Manchester

UNITED STATES - ÉTATS-UNIS
OECD Publications and Information Centre,
2001 L Street, N.W., Suite 700,
Washington, D.C. 20036 - 4095
Tel. (202) 785.6323

VENEZUELA
Libreria del Este,
Avda F. Miranda 52, Aptdo. 60337,
Edificio Galipan, Caracas 106
Tel. 951.17.05/951.23.07/951.12.97

YUGOSLAVIA - YOUGOSLAVIE
Jugoslovenska Knjiga, Knez Mihajlova 2,
P.O.B. 36, Beograd Tel. 621.992

Orders and inquiries from countries where Distributors have not yet been appointed should be sent to:
OECD, Publications Service, 2, rue André-Pascal,
75775 PARIS CEDEX 16.

Les commandes provenant de pays où l'OCDE n'a pas encore désigné de distributeur doivent être adressées à :
OCDE, Service des Publications. 2, rue André-Pascal, 75775 PARIS CEDEX 16.

71784-07-1988

OECD PUBLICATIONS, 2, rue André-Pascal, 75775 PARIS CEDEX 16 - No. 44561 1988
PRINTED IN FRANCE
(97 88 06 1) ISBN 92-64-13151-5